English Urban Commons

This book presents a novel examination of urban commons which provides a robust base for education initiatives and future public policy guidance on the protection and use of urban commons as invaluable urban green spaces that offer a diverse cultural and ecological resource for future communities.

This book's central argument is that only through a deep understanding of the past and a rigorous engagement with present users can we devise new futures or imaginaries of culture, well-being and diversity for the urban commons. It argues that understanding the genesis of, and interactions between, the different pressures on urban green space has important policy implications for the delivery of nature conservation, recreational access and other land use priorities. The stakeholders in today's urban commons, whether land users, policy makers or the public, are the inheritors of a complex cultural legacy and must negotiate diverse and sometimes conflicting objectives in their pursuit of a potentially unifying goal: a secure future for our urban commons. This book offers a unique and strongly interdisciplinary study of urban commons, one that brings together original historical investigation, contemporary legal scholarship, extensive oral history research with user groups and research examining the imagined futures for the urban common in modern society. It explores the complex social and political history of the urban common, as well as its legal and cultural status today, using four diverse case studies from within England as exemplars of the distinctively urban common. These are Town Moor in Newcastle, Mousehold Heath in Norwich, Clifton and Durdham Downs in Bristol and Valley Gardens in Brighton. This book concludes by looking forward and considering new tools and methods of negotiation, inclusivity and creativity to inform the future of these case studies and of urban commons more widely.

This book will be of great interest to students and scholars of the commons, green spaces, urban planning, environmental and urban geography, environmental studies and natural resource management.

Christopher Rodgers is an Emeritus Professor of Law at Newcastle University, UK.

Rachel Hammersley is Professor of Intellectual History in the School of History, Classics and Archaeology at Newcastle University, UK.

Alessandro Zambelli is a Reader in Architecture at the University of Portsmouth, UK.

Emma Cheatle is Senior Lecturer in Architecture at the University of Sheffield, UK.

John Clarke is an Associate Professor of Creative Writing and Academic Director for Arts and Culture at the University of Exeter, UK.

Sarah Collins is a Research Associate in History in the School of History, Classics and Archaeology at Newcastle University, UK.

Olivia Dee was a Research Associate in Oral History at Newcastle University. She is now a Research Fellow at Queen's University Belfast, UK.

Siobhan O'Neill is an applied performance practitioner and a Research Fellow at Portsmouth School of Architecture, UK.

Earthscan Studies in Natural Resource Management

Balancing the Commons in Switzerland
Institutional Transformations and Sustainable Innovations
Edited by Tobias Haller, Karina Liechti, Martin Stuber, François-Xavier Viallon and Rahel Wunderli

Natural Resource Sovereignty and the Right to Development in Africa
Edited by Carol Chi Ngang and Serges Djoyou Kamga

Reindeer Husbandry and Global Environmental Change
Pastoralism in Fennoscandia
Edited by Tim Horstkotte, Øystein Holand, Jouko Kumpula and Jon Moen

The bioeconomy and non-timber forest products
Edited by Carsten Smith-Hall and James Chamberlain

Drylands Facing Change
Interventions, Investments and Identities
Edited by Angela Kronenburg García, Tobias Haller, Han van Dijk, Cyrus Samimi, and Jeroen Warner

Blue Flag Beaches
Economic Growth, Tourism and Sustainable Management
Edited by María A. Prats and Fernando Merino

Resource Communities
Past Legacies and Future Pathways
Kristof Van Assche, Monica Gruezmacher, Lochner Marais, and Xaquin Sindin

English Urban Commons
The Past, Present and Future of Green Spaces
Christopher Rodgers, Rachel Hammersley, Alessandro Zambelli, Emma Cheatle, John Clarke, Sarah Collins, Olivia Dee and Siobhan O'Neill

For more information about this series, please visit: www.routledge.com/books/series/ECNRM/

English Urban Commons

The Past, Present and Future of Green Spaces

Christopher Rodgers, Rachel Hammersley, Alessandro Zambelli, Emma Cheatle, John Clarke, Sarah Collins, Olivia Dee and Siobhan O'Neill

Routledge
Taylor & Francis Group
LONDON AND NEW YORK

earthscan
from Routledge

First published 2024
by Routledge
4 Park Square, Milton Park, Abingdon, Oxon OX14 4RN

and by Routledge
605 Third Avenue, New York, NY 10158

Routledge is an imprint of the Taylor & Francis Group, an informa business

Funded by Newcastle University.

British Library Cataloguing-in-Publication Data
A catalogue record for this book is available from the British Library

Library of Congress Cataloging-in-Publication Data
Names: Rodgers, C. P. (Christopher P.), author. | Hammersley, Rachel, 1974–author. | Collins, Sarah (Historian), author. | Dee, Olivia, author. | Zambelli, Alessandro, author. | O'Neill, Siobhan, author. | Cheatle, Emma, author. | Clarke, John Wedgwood, author.
Title: English urban commons : the past, present and future of green spaces / Christopher Rodgers, Rachel Hammersley, Alessandro Zambelli, Emma Cheatle, John Wedgwood Clarke, Sarah Collins, Olivia Dee, and Siobhan ONeill.
Other titles: Past, present and future of green spaces | Earthscan studies in natural resource management.
Description: New York, NY : Routledge, 2024. | Series: Earthscan studies in natural resource management | Includes bibliographical references and index.
Identifiers: LCCN 2023027091 (print) | LCCN 2023027092 (ebook) | ISBN 9781032069180 (hardback) | ISBN 9781032069210 (paperback) | ISBN 9781003204558 (ebook)
Subjects: LCSH: Commons—Great Britain. | Commons—England. | Sociology, Urban—Great Britain.
Classification: LCC HD1289.G7 R63 2024 (print) | LCC HD1289.G7 (ebook) | DDC 333.20941—dc23/eng/20230922
LC record available at https://lccn.loc.gov/2023027091
LC ebook record available at https://lccn.loc.gov/2023027092

ISBN: 978-1-032-06918-0 (hbk)
ISBN: 978-1-032-06921-0 (pbk)
ISBN: 978-1-003-20455-8 (ebk)

DOI: 10.4324/9781003204558

Typeset in Times New Roman
by codeMantra

Contents

Preface

This book presents research arising from an interdisciplinary research project "Wastes and Strays: The Past, Present, and Future of English Urban Commons", which was funded by the Arts and Humanities Research Council between 2019 and 2022. It makes a compelling case for"urban" commons as publicly accountable and open green spaces vital for culture, health, well-being and biodiversity. This project explored the unique contribution that urban commons have made – and could in future make – to community well-being and the urban environment in our towns and cities. Whatever their legal status, urban commons have always reflected shifts in social and cultural attitudes and traditions and have long been sites of debate and negotiation. Their future in the contemporary metropolitan context is, however, unclear. The COVID-19 pandemic and associated restrictions on work and travel have illustrated just how important urban green space is for recreational use, and emphasised the great physical and mental health benefits that access to our urban commons generates. This book explores how we can champion our urban commons and explores new methodologies to engage the public in their protection, use and development as invaluable community resources for the future. A central theme that emerges is that only through a deep understanding of the past, and rigorous engagement with present users, can we devise future plans for our urban commons that reflect and respond to the needs of multiple user groups and allow for a range of overlapping uses.

This book seeks to offer a unique and strongly interdisciplinary study of urban commons, one that brings together original historical investigation, contemporary legal scholarship, extensive oral history research with commons user groups and participatory research examining the imagined futures for the urban common in modern society. It explores the complex social and political history of the "urban" "common", as well as its legal and cultural status today, using four diverse case studies as exemplars of the distinctively "urban" common. These are Town Moor (Newcastle), Mousehold Heath (Norwich), Clifton and Durdham Downs (Bristol) and Valley Gardens (Brighton). This book also looks to the future and considers new tools and methods of negotiation, inclusivity and creativity to inform the future of these case studies and of urban commons more widely. The historical research and legal scholarship undertaken by the project was complemented by an extensive programme of fieldwork that involved (i) oral history fieldwork with current commons users and user groups and (ii) participatory action research with commons users. The participatory action research undertaken for the project emphasised collaborative approaches in which knowledge and visions of our case study commons was co-produced with local people; accordingly, participants were given the opportunity to be credited by name for their contributions. The interview transcripts from the oral history research have, on the other hand, been coded to preserve the anonymity of participants. It follows that the referencing for the respective fieldwork sources is presented differently in the text, and especially in the case study chapters (Chapters 2–5).

In presenting the findings of the Wastes and Strays research programme, the authors wish to acknowledge the financial support of the Arts and Humanities Research Council, without which this research would not have been possible. We are also grateful to all the participants in our fieldwork and to the members of the project's Advisory Board. We would like, in particular, to record our thanks to the late Cllr. Nigel Todd (Newcastle), who gave valuable support and assistance in the early development of the project, but sadly passed away before it came to fruition. Nigel will be sadly missed, but we very much hope he would have approved of our subsequent efforts to champion our urban green spaces.

The project was impacted by the COVID-19 pandemic, which necessitated changes and delays to the project fieldwork programme and additional funding from the research council. The project concluded with four exhibitions of our research findings, held in each of the case study locations, and we are grateful for the support of the Wylam Brewery (Newcastle), Frere Road Community Centre (Norwich), Clifton Observatory (Bristol) and the Open Market (Brighton) in hosting exhibitions of our work in May and June 2022. It is hoped that the project exhibition will become a permanent online exhibition. Further information about the project, including webinars and a dedicated blog programme, can be found at our website https://research.ncl.ac.uk/wastesandstrays/.

Christopher P. Rodgers
Rachel Hammersley
Alex Zambelli

1 The Urban Common as "Common" Space

Urban Commons in Context

Why are our urban commons so "special"? What singles them out from other categories of urban "green space" as worthy of special study, protection and attention? They provide a common space for public recreation, for preserving and promoting biodiversity, and have (and still are) sites with distinct histories and special qualities that engender artistic and poetic invention and engagement. These features also distinguish urban commons in the public mind from other categories of public space – for example, town and village greens, or public parks. What are these features, and why do they resonate so powerfully in the public mind? These are all questions that this work will seek to interrogate and explore.

Urban commons provide invaluable, open, public space in dense urban settings where such space is at a premium. As complex, wide-ranging ecosystems, they accommodate various recreational uses, including sports and leisure pastimes, and host important flora and fauna. They are often less regulated spaces, allowing "freedom", diversity and the sense of the unexpected and the imaginary within the city, yet are often underused and undervalued by local communities and at risk to commercial development. In the common consciousness, the urban common is often understood as "a green space owned by all". However, each urban common has a different origin and legal status, as well as a series of particular uses and perceptions that are grounded in, but not always determined by, those of the past.

Each urban common is an open space defined by the complexity of its origins, land use characteristics and cultural intersections with the history and significance of "place" of its particular urban setting. Indeed, when studying the past, present and future of our urban commons, it is important to remember the dichotomy between an urban common as a green *space* and as a *place* of historical and communal significance. As a "place", it is best understood in a static sense – somewhere where elements are organised in relations of co-existence. Whereas the notion of a common as a "space" for community use is reflective of the influence upon it of mobile elements or "the ensemble of movements deployed within it…..[in other words] *space is a practiced place*".[1] The notion of an urban common as a green space is therefore dynamic and reflects the uses made of it – for example, as recreational space or for promoting community gatherings for protest or sports. A central

DOI: 10.4324/9781003204558-1

aim of this book is to bring together an understanding of our urban commons in both senses – and in doing so, to promote a better understanding of their origins and characteristics in order both to protect and enhance them and to promote their use as commons for future generations.

In a post-pandemic age of austerity, the protection and promotion of urban green space, for sustainable well-being in diverse city communities, is critical. Some urban commons are registered common land, with their legal status protected by law; others are safeguarded by specific Acts of Parliament; while yet others are spaces without any formal legal or planning recognition, in which case they will have a very different and less securely protected legal status. If urban commons are to be protected for future generations, new and secure urban commons created, their value enhanced and new and imaginative uses of them encouraged, then a proper understanding of them grounded in historical and legal research and informed by current perceptions and use is vital. This book presents the outcomes of interdisciplinary research intended to (i) develop a robust and multifaceted definition of what constitutes an "urban common", with practical utility for a wide range of user groups and stakeholders, and (ii) develop new and imaginative modes of public engagement that can protect, champion and promote the use of urban commons by local communities. It is hoped, therefore, that the outcomes presented in this work will provide a new framework for thinking about the history, cultural significance, protection and future use of these undervalued and poorly understood spaces.

The following chapters will address these overarching questions using a case study methodology, focused on four diverse case studies of important surviving urban commons presented in Chapters 2–5: Town Moor, Newcastle upon Tyne (Chapter 2); Mousehold Heath, Norwich (Chapter 3); Clifton and Durdham Downs, Bristol (Chapter 4); and Valley Gardens, Brighton (Chapter 5). These will also provide a secure foundation and a model for investigating key characteristics of *all* urban commons in England and Wales, and these will be explored in the thematic chapters (6–10) following the case studies.[2]

Concepts of Space and the "Common" as Urban Space

The central argument emerging from the case studies and accompanying analysis of contemporary urban commons in this work is that in order to understand their role and place in their modern social setting, it is important to understand their past – their origins, how these interlink with their legal status at different points in their history and how the uses to which each common is (and has) been put shape and determine contemporary notions of "belonging", local identity and a community's vested interest in the land as a common space. These interlinked strands are explored in this work using an interdisciplinary mixture of historical methods; legal scholarship; oral history research with contemporary users and commons user groups; and participatory and performative research with users to identify future imaginations of each of the urban commons as a shared and valued community space.

The case studies of contemporary urban commons presented in Chapters 2–5 reveal four areas of communally used "space" that are very different; different in their origin; with different histories and contemporary uses as a community resource; different in the way in which their contemporary management is organised and delivered; and with a very different formal legal status in each case. But for all their differences, the case studies also illustrate several key facets of *all of our* contemporary urban commons as a social phenomenon: each is uniquely important in providing recreational space for their communities, and each plays an important role in defining locality and identity in its local context.

Understanding the interplay between these different factors and how they define what is (and what is not) an "urban common" requires a consideration of wider concepts of "space" and community, how urban space is created as a social construct and how the urban "common" as a distinct type can be positioned within broader notions of communal space. In this context, Henri Lefebvre's theoretical framework[3] for analysing spatial concepts is helpful. Lefebvre suggests we focus on three different (but related) aspects of common space and consider the dialectical relationship that exists between them. His triad requires us to distinguish between "conceived" space, "perceived" space and imagined (or "representational") space.[4] The key insight this facilitates is that the spatial practice of a society propounds and presupposes space in a dialectical interaction. It produces it slowly – it is revealed through its deciphering of its space (what Lefebvre calls "perceived" space). This will usually be embodied in the close association between daily routines and urban reality – for example, the routes and networks that link up places set aside for work, private life and leisure. This must be contrasted with the conceived aspects of space; that is, the conceptualised space of scientists or planners, all of whom identify what is lived and what is perceived with what is conceived ("conceived" space). This is the dominant space in any society and reflects, for example, the pre-existing legal order for the regulation and use of common space. Conceptions of space tend, with some exceptions, towards a system of verbal signs. The third element integral to notions of space is "representational" space – space as directly lived through its associations with images and symbols and hence the space of inhabitants and users. Lefebvre considers this to be the passively experienced space which the imagination seeks to change and appropriate ("lived" space). The triad is especially helpful in framing the interactions between historical, contemporary and future uses of urban commons as a special form of urban green space – conceived space is largely concerned with the past (e.g. the legal and property framework within which the common has developed over past centuries), perceived space with contemporary uses and identifications of (and with) the common space and representational space with the future aspirations of the community for the space.

Separating these interlinked elements of common space as a social construct provides a powerful analytical tool when thinking about the past, present and future of our urban commons. In the first instance, we need to separate out factual accounts – for example, those legal rules, norms and structures governing the status and management of an urban common – from a consideration of the common as a socially created space. The latter is addressed in this work through, for example, the use of oral history research to examine the "creation" of space in the understanding of commons users and user groups. Moreover, the legal constructs that in theory govern the ownership, use and management of our urban commons are not themselves always grounded either in actual history or in the social reality of the commons' contemporary use. Furthermore, where their modern status and legal characterisation has been dictated by historical fact this quickly becomes obsolete, with the consequence that the conceived nature of an urban common as common "space", one that is reflected in its legal status, is not representative of the social reality in a modern context. The contemporary hierarchy of legal norms that in theory govern and shape issues of ownership and land use do not, in many cases, reflect historical reality. A fundamental example of this lies in the agricultural origins of many of the user "rights" still subsisting over some of our urban commons. Many contemporary urban commons were originally the wastes of a manor over which manorial tenants had rights to graze livestock and exercised other land use rights, including estovers (the right to take fallen wood and other natural produce of the land) and turbary (the right to cut peat and turf).[5] In other words, they were predominantly rural and agricultural in character. Where the social context for these commons is now predominantly urban, these former agricultural resource use rights often still subsist even if they are not, in a modern context, either exercised or relevant. The registration under the Commons Registration Act 1965 of grazing and other rights over what are today predominantly urban commons is a good example of this phenomenon.[6]

Past, Present and Future

Interpreting the relevance and importance of the history of our urban commons in shaping their contemporary and future use is a multifaceted and potentially complex exercise. One way to approach this task is to distinguish between historical *events* and *structures*. All events are based on pre-existing structures that become part of the events themselves, but that existed before the events in a different way from the chronological sense of the "before".[7] In this context, the legal rules that in theory govern the ownership and use of an urban common themselves shape the historical events that are important to an understanding of its current or future status and use. The strength of customs and legal rules tends to arrange and outlive individual events or acts.[8] We will consider concepts of "ownership" and "belonging" in the case studies presented in this work and the relationship that these have to the legal and historical framework within which each urban common must be considered. Urban green "space" is a social construction, but the way in which this is constructed in each case will depend upon the interaction between the structures that have governed its use and development and the events relevant to an understanding of its history and importance to the local community.

Put simply, the characterisation of a common by contemporary commons users and user groups as "belonging" to the community, or its status in the public's mind as a "unique" social space of communal importance, will be shaped and coloured by the historical structures within which that common has developed over time, for example, its specific and unique legal status, the customs that have arisen and that govern its use by individuals and by the wider community. The "horizon of expectation" of commons users and their lived experience will each interact and mutually inform each other and their development, joining together the past and the future in an asymmetrical manner.[9] Oral history research with contemporary commons users and user groups can inform and develop a narrative exploring the meaning and significance of an urban common as community "green space". But this narrative will itself have been influenced and shaped by (i) the historical structures that are in place to govern the use and management of the common (e.g. its legal status and the different property rights recognised as subsisting over it and how they have been used) and (ii) the role that the common has played in the past as a site of contestation, of protest, or as a communal meeting place or a site for public recreation.

Key Definitions: Commons, Common Rights and Urban Commons

Defining a "Common"

An urban common is, in legal terms, simply a particular type of common to which special rules apply, for example, regarding the making and implementation of management schemes governing its use. This means that we must first consider the more general definition of a common before we can explore the contemporary understanding of what constitutes an "urban" common. There is no single, overarching definition of what constitutes a common in English Law, but the key characteristics of commons have their origins in agricultural land management under the medieval manorial system of land use. This perhaps surprising fact explains many of the anomalies manifested by the contemporary governance of urban commons – whose functions and utility to local communities reflect the realities of a very different social setting to that from which the defining characteristics of a "common" are derived.

Though the notion of "commons" today invokes the idea of a wide and inclusive ownership, common rights were traditionally exclusive and restricted. A common is a piece of land owned by one person (traditionally the lord of the manor) but over which other people had certain defined common rights. Until 1925, the public had no established rights over common land; rather, these were held by individual commoners and were usually attached to specific tenancies or dwellings. In *Halsbury's Laws of England*, a common right is defined as "a right, which one or more persons may have, to take or use some portion of that which another person's soil naturally produces".[10] In legal terms, it is a *profit a prendre*,[11] but it is one held "in common" because the holder of the right shares the right to take the produce of the land with the owner of the soil.[12] The commonality referred to is therefore that between the commoner (the holder of the right to take the produce of the land) and the owner of the land itself – it has nothing to do with communal or community rights or land use. Traditionally, there are six types of common right.

By far, the most typical is the right of *pasturage* or *grazing*. This was usually organised either by *levancy* and *couchancy* or by stints. The former required that the livestock grazed on the common were no more than could be supported on winter herbage and hay of the tenement to which the rights were attached. Though rights were attached to particular tenements, the exercise of them often incurred a fee.

This was used to pay officials who were appointed to look after the land and care for the stock and sometimes also to provide a fund to support commoners and their widows and orphans in times of hardship. In some places, grazing was prohibited between Candlemas (2 February) and Lammas Day (12 August until 1762, 1 August thereafter) to allow for the production of hay for the wintertime. Other common rights include the right of *estover*, typically a right to collect small amounts of wood, furze or bracken and usually encountered when commons included woodland or heath. In woodland areas, the right of *pannage* allows the owners to feed pigs on fallen acorns, beechmast, chestnuts and nuts. *Piscary* is the right to take fish from waters that belong to someone else. The right of *turbary* allows the holder to cut turf or peat for fuel or for roofing and walls. Finally, the right of *common in the soil* is the right to take natural resources such as sand, gravel, stones or minerals. These common rights, and their contemporary relevance (or otherwise) for the governance of our urban commons, are discussed more fully in Chapter 8.

While the landowner enjoyed certain rights over the land (including usually the residue of the natural products of the land once the commoners' rights had been satisfied), s/he could not interfere with the exercise of those rights in any way. The dual nature of this system produced an in-built tension which has often been replicated in later legislation.

Defining an "Urban" Common

It is, perhaps, surprising that the "urban" common has received less attention and research than commons and common land more generally. Much scholarly attention has been devoted in recent years to the study of our common lands, both in the UK and elsewhere,[13] but little attention has been devoted to the specifically "urban" commons as opposed to commons in rural areas or the "commons" more widely conceived as a social construct of public space.[14] Henry French has suggested that one reason for this scholarly neglect may be that commons are generally "regarded as a phenomenon of rural society" due to their association with agricultural activity.[15]

Yet, town or urban commons were far from unusual; indeed, most historic towns had at least one common.[16] Their main original function was to provide grazing for the draft animals required by local artisans and merchants. Consequently, they have been described as the precursor of the city car park.[17] But urban commons provided other benefits too. Part of the appeal of Newcastle's Town Moor was the provision of cheap milk for the city's inhabitants, and both it and Bristol's Downs were the site of reservoirs that provided their respective cities with water. Moreover, commons were spaces on which large numbers of people could gather for the purposes of politics, religion or entertainment. And, like their rural counterparts, they could also provide a crucial safety net for locals during hard times, both by the income generated via fees from grazing or temporary enclosures and from the exercise of the right to gather building materials or fuel when required.

What then is an "urban" common? And what distinguishes it in a modern context from other types of space recognised as "commons"? One anomaly of the legislation governing urban commons is that the definition of what constitutes an "urban" common is largely *referential* – urban commons are usually defined by reference to whether they are within the geographical boundaries of an urban local government area. The Metropolitan Commons Act 1866, for example, applies to commons within the metropolitan police district[18] and the generally applicable rights of access for air and recreation apply to commons which are situated in an area which before 1 April 1974 was a borough or urban district or is within the metropolis within the meaning of the 1866 Act.[19]

There is no *functional* definition of an urban common in the current legislation governing their management and use. The legislation provides for schemes of management to be made for each urban common, as shall be explained below,[20] and these reflect in each case their unique features and contribution in their specific urban setting. But the very different functions provided today by urban commons – as opposed to commons more generally – is not captured in the legal definition of an urban common. Indeed, they must possess the wider characteristics of common land as a prerequisite and also be situated in the metropolitan area or an urban district if they are to come within the provisions of the legislation on urban commons. This means that the requirement for them to be former wastes of a manor, or to have subsisting common rights over them, remains relevant to their categorisation as urban *commons*,[21] even though these factors have no relevance to their utility and social function in a modern context.

Henry French has identified five distinct types of urban or town common:[22]

(1) Those in which grazing rights were exercised by freemen over lands within or adjoining the borough or township which were sometimes owned by neighbouring manorial lords or landowners. These rights were often restricted to certain times of the year.

(2) Lowland arable land which was owned by the lord(s) of the manor or corporation but over which manorial tenants or borough freemen – or in practice, sometimes all rate-paying inhabitants – had rights.

(3) Pasture land within the boundaries of the township or borough that was owned by the lord(s) of the manor or corporation. This might comprise both land that could be grazed after harvest or when fallow and dedicated pasture commons.

(4) A sub-variant of three – inland commons of upland townships within the boundaries of the settlement. These could be arable or pasture.

(5) Upland grazing rights for town residents over nearby moorland.

Some commons were always urban in character, designed to serve the city population. This was the case with Newcastle's Town Moor. Others were originally more rural – separated from the city by agricultural land. These became "urban" commons as a result of urban development and expansion. This was what happened with Bristol's Clifton and Durdham Downs.

The History of Urban Commons

The precise origins of English common land are obscure. Most charters establishing common rights date from the late medieval period, and it has been argued that common rights arose at this time when more land was being brought into cultivation leaving less "wasteland". But there is also evidence of common land dating back to Anglo-Saxon times, and the late medieval charters probably confirmed rights that had already been enjoyed for some time.[23]

Over the centuries, there has been a significant reduction in the extent of surviving common land in England, and urban commons have been particularly affected.[24] Enclosure of common land was made possible by the Statute of Merton of 1235 which declared that enclosure was legitimate as long as sufficient land was left for the commoners. The first period of enclosure, which lasted from the mid-fourteenth to the early eighteenth century, primarily involved the enclosure of agrarian land for agricultural purposes. Urban commons were most directly affected during the second phase (c.1750–1845), which is known as the period of parliamentary enclosure. The 1845 Enclosure Act did not bring an immediate end to the practice of enclosure,[25] but it did initiate a more formal process and required compensation to be provided when land was enclosed. Regulations were tightened further during the second half of the nineteenth century. The Metropolitan Commons Act 1866 prohibited the enclosure of common land within the Metropolitan Police District. The Commons Act 1876 required "benefit to the neighbourhood" to be demonstrated when an application for enclosure was made. As a result of these tighter regulations, by the end of the century, enclosure had largely ceased. The Commons Act 1899 reflected the new emphasis on the regulation and management (rather than the enclosure) of common land. This was further secured in the twentieth century with the Commons Registration Act 1965, which was designed to preserve open spaces and to protect them from commercial development. It has been argued, however, that the definition of common land that was employed in the Act, and the procedure for registration, were at odds with traditional understandings and practices, thereby resulting in a number of "misregistrations".[26]

The uses to which urban commons were put also changed over time.[27] In medieval times, even urban commons were largely used for agricultural purposes, though there is some evidence of coal mining on Newcastle's Town Moor at this date. During the early modern period, industrial uses increased with mining, quarrying, the erection of windmills and the construction of reservoirs occurring with greater frequency in these spaces. From the eighteenth century, recreation gradually took over as a major – often the dominant – use of urban commons, with both informal pursuits such as walking and riding and more formal organised sports becoming popular. This shift was finally recognised in the Law of Property Act 1925, which gave the public "rights of access for air and exercise" over all urban and metropolitan commons.[28] Today, recreation remains the dominant use of these spaces, though they are also increasingly seen as sites that have the potential to bring environmental benefits.

A "Tragedy" of the Urban Commons?

As we have seen, many urban commons in England and Wales today ultimately derived their contemporary legal identity and property structures from the manorial system. Once important sources of natural resources for local communities and integral to the agricultural management of the land, their social functions and cultural identity has fundamentally changed over the centuries. With a few exceptions (e.g. Town Moor, Newcastle), they no longer provide natural resources such as grazing for livestock and are instead primarily seen as spaces for recreation and a multitude of other community activities that use what is essentially "shared" space in an urban setting. Nevertheless, their legal categorisation as a type of common land, albeit a very special type, raises the question whether they are, like other species of communal resource, susceptible to the "tragedy of the commons" famously identified by Garrett Hardin.[29] Hardin identified the inherent tension between individual self-interest and the communal good as a fundamental flaw in the management of common resources, a conflict that would lead inevitably to the "tragedy of the commons" and the destruction of the resource in question. The principal challenge in this scenario is "free riding", leading to the exhaustion of the commonly shared resource and how best to address it.

The tragedy of the commons thesis has been extensively challenged in subsequent literature, most notably by the work of Elinor Ostrom, which has demonstrated that sustainable management of common resources is possible.[30] Nevertheless, there remains an inherent conflict between individual and communal interests in common land.[31] No individual can take sole control or dispose of the resource. The property rights regime on common land contained within it a set of tensions between the rights of the owner of the soil and those having communal land use rights. The owner of the soil cannot exercise unfettered control and use of the land, as s/he cannot interfere with or restrict the rights of commoners to take produce or of the public to open access for recreational enjoyment. But the interests and actions of everyone are potentially destructive and must be carefully balanced by the community of users if the common is to be sustained for future use.

The tragedy of the commons debate poses problems when we consider its relevance (or otherwise) to urban studies and to the management of urban commons in particular. Both Hardin and Ostrom define the "commons" as a common pool resource (CPR), a concept that includes most natural resources such as grazing lands, fisheries and forests.[32] As a consequence, the commons are depicted as a "self-evident resource (object) that only waits for its appropriators (subject) to exploit it".[33] The problem with this analysis when transposed to the urban setting is that the act of consumption does not *detract* but rather *increases* the value of urban resources to the community, and this is especially the case with public parks and urban commons. In the context of urban commons, value is a relational concept that is dependent on, for instance, factors such as the proximity of the green space to the community and the density of population in the urban area concerned. It can therefore be argued that usage and consumption are a constitutive part of the production of the urban commons.[34] The urban commons are essentially dynamic; particular urban commons are not simply resources waiting to be exploited but must first be produced and then constantly reproduced.[35] In other words, space is constituted and given meaning through human endeavour – it is ontogenetic in the sense that it is brought into existence through transductive practices that change the conditions under which the space is (re)made.[36]

This understanding of an urban common has resonance in the context of the work presented in this study: an urban common can only be developed and maximised if it is used, and its characterisation, importance and significance to the local community will be shaped and determined by the types of use to which it is put. The "use" is what makes it a "common". Chapters 9 and 10 explore the development of future imaginations of the urban common, drawing on participatory research with commons user groups undertaken in our case study commons, and consider strategies for their future development as participatory communal spaces. It follows that instead of thinking about the "tragedy of the [urban] commons" in terms of resource diminution,[37] it is more productive to think of it in terms of a "tragedy of urban commoners", viz. the "tragedy" that occurs when citizens are not able to pursue their individual interests in the public space represented by our urban commons. The promotion of "commoning" as a social practice is a key objective of this study.

Book Structure

The research on which this book is based was carried out as part of an AHRC-funded project: "Wastes and Strays: The Past, Present, and Future of Urban Commons". Researchers working in various disciplines (law, history, architecture, literature) at four different institutions engaged in archival research, desk-based archaeological work and field surveys, the collection of oral histories and community engagement. Work centred on four contrasting, yet representative, case studies of surviving urban commons: Town Moor, Newcastle upon Tyne; Mousehold Heath, Norwich; Clifton and Durdham Downs, Bristol; and Valley Gardens, Brighton. This approach is reflected in the structure of this book.

Part One, comprising Chapters 2–5, offers overviews of the four case studies. Each one covers the changing space of that particular common, the property rights and governance structure associated with it, the place of that common in its historical context, its contemporary use and its imagined future.

Part Two adopts a more thematic approach and develops an argument that links the past, present and future of England's urban commons. Chapter 6 explores the themes of continuity and change in the shape, land use and culture of urban commons. It argues that while these spaces might appear to be vestiges of the past in the present, they have, in fact, always been dynamic and adaptable spaces that have changed their shape, appearance and use over time. Focusing on the themes of conflict and contestation, Chapter 7 demonstrates that as well as changing over time, urban commons have historically been amenable to a variety of understandings at a single point in time. The particular nature of these spaces has encouraged the development of diverse, sometimes even incompatible, visions of how they should look and the appropriate uses to which they can be put. These competing visions have repeatedly led to conflict. Moreover, these conflicts have frequently been exacerbated by the legal methods that have been used to protect, preserve and manage these spaces. Legal mechanisms, such as Acts of Parliament, have tended to be concerned with offering a solution to specific problems at a particular point in time.

As a result, they generally favour one vision of the urban common rather than recognising a multiplicity of uses, and they provide a fixed and inflexible solution that quickly becomes outdated and unworkable as times change. Building on this history, Chapter 8 acknowledges that historical models of commons governance are no longer appropriate for the management of urban commons in the twenty-first century and argues that in order to better protect and enhance the use value of urban commons in the future, we need to develop a new approach to commons governance, one based on identifying and then maximining the ecosystems services that they can provide for the community.

Notes

1 See Michel de Certeau, *The Practice of Everyday Life*, (Berkeley: University of California Press, 1984), p. 117.
2 The social, legal and historical context for urban commons in Scotland is very different to that in England and Wales and is not explored in this work. See for example Angus Winchester, *Common Land in Britain: A History from the Middle Ages to the Present Day* (Woodbridge, NJ: The Boydell Press, 2022), pp. 29–31, 44–45.
3 See Henri Lefebvre, *The Production of Space* (Oxford: Blackwell tr. Donald Nicholson-Smith, 1991).
4 Ibid., p. 38.
5 See further Chapter 8, p. 181 below.
6 See for example the user rights registered over Durdham common in Bristol, discussed in Chapter 4 at p. 100.
7 See Reinhart Koselleck, *The Practice of Conceptual History, Timing History, Spacing Concepts* (tr.TS Presner, Stanford: Stanford University Press, 2002), esp. at p. 123ff.
8 Ibid., p. 125.
9 Ibid., p. 127
10 Halsbury's Laws of England, Commons (Volume 13 (2021)) 1. Introduction to Commons (2) Definitions Relating to Commons 305. Meaning of 'right of common'. https://plus.lexis.com/uk/document/documentlink/?pdmfid=1001073&crid=59750fc8-1aa9-42fa-995f-ca2c969f3c2a&pddocfullpath=%2Fshared%2Fdocument%2Fanalytical-materials-uk%2Furn%3AcontentItem%3A8VFC-K4H2-D6MY-P54B-00000-00&pdcontentcomponentid=127805&pdproductcontenttypeid=urn%3Apct%3A545&pdiskwicview=false&pdpinpoint=_0089_1_Commons_HTENCY-SUBJECT_305_HTENCY-PARA&ecomp=-k4k&prid=1deaf42b-c1d7-4c67-954c-305bcd75bd45#.
11 *Beckett (Alfred F) Ltd v Lyons* [1967] Ch 449 at 482 per Winn LJ.
12 See generally, Richard Honey and Edward Cousins, *Gadsden and Cousins on Commons and Greens* (3rd ed., Mytholmroyd: Sweet & Maxwell, 2020), pp. 2-05–2-07.
13 See for example: Angus Winchester, *Common Land in Britain: A History from the Middle Ages to the Present Day* (Martlesham: The Boydell Press, 2022); Christopher Rodgers, Angus Winchester, Eleanor Straughton, and Margherita Pieraccini, *Contested Common Land: Environmental Governance Past and Present* (London: Earthscan Publishing, 2011).
14 For example: David Harvey, *Social Justice and the City* (London: Edward Arnold, 1973); David Harvey, *Spaces of Hope* (Oxford: Blackwell, 2000); Mark Bowden, Graham Brown, and Nicky Smith, *An Archaeology of Town Commons in England 'A Very Fair Field Indeed'* (Swindon: English Heritage, 2009), p. vii; John P. McCarthy and Michael E. Leary-Owhin, 'Representing and Contesting Urban Space', Part 3 Chapters 197–280, in *The Routledge Handbook of Henri Lefebvre, The City and Urban Society* (London: Routledge, 2020).

15 Henry French, 'Urban agriculture, commons and commoners in the seventeenth and eighteenth centuries: the case of Sudbury, Suffolk', *The Agricultural History Review* 48:2 (2001): 172.
16 Bowden, Brown, and Smith, *An Archaeology of Town Commons in England*, p. 1.
17 Ibid., p. 1.
18 Metropolitan Commons Act 1866, s.4 and sched.1.
19 Law of Property Act 1925, s. 193. Discussed further below, Chapter 8, p. xxx.
20 See Chapter 8, p. 186ff.
21 See for example Metropolitan Commons Act 1866, s.3.
22 French, 'Urban agriculture, commons and commoners in the seventeenth and eighteenth centuries', 173–176.
23 Bowden, Brown and Smith, op.cit., pp. 16–18.
24 For a more detailed discussions of the history of enclosure see Chapter 6 and for more on the legislation on common land, see Chapter 8.
25 Alun Howkins, 'The Use and Abuse of the English Commons 1845-1914', *History Workshop Journal* 78 (2014): 107–132.
26 Sara Birtles, 'The Impact of Commons Registration: A Norfolk Study', *Landscape History* 20:1 (1998): 83–97.
27 Once again, what is offered here is merely a broad sketch. A more detailed account is provided in Chapter 6.
28 Law of Property Act 1925, s.193.
29 See G. Hardin, 'The Tragedy of the Commons', *Science* 162 (1968): 1243–1248.
30 See Elinor Ostrom, *Governing the Commons: The Evolution of Institutions for Collective Action* (Cambridge: Cambridge University Press, 1990).
31 See further Rodgers, Winchester, Straughton, and Pieraccini, op.cit., pp. 9–16.
32 See Ostrom, op.cit., p. 413.
33 Christian Borch and Martin Kornberger, *Urban Commons – Rethinking the City* (London: Routledge, 2015), p. 6.
34 See for example Christian Borch and Martin Kornberger, op.cit., p. 7.
35 See Leif Jerram, Chapter 2, pp. 47–67 in Christian Borch and Martin Kornberger, op.cit.
36 Martin Dodge and Rob Kitchin, 'Code and The Transduction of Space', *Annals of the Association of American Geographers* 95:1 (2005): 162–180, see esp. 171ff.
37 We should, for example, think of the urban common not as a "resource" but as a malleable social relation: see David Harvey, *Rebel Cities: From the Right to the City to the Urban Revolution* (New York and London: Verso, 2012), p. 73.

Part I

The English Urban Common

Four Case Studies

2 Town Moor, Newcastle upon Tyne

Introduction

Location

Newcastle's urban common covers an area of 388.15 hectares to the north and west of the city centre.[1] It consists of two larger sections, Nuns Moor in the north-west and the central Town Moor (a label that is also used to refer to the whole area), together with four smaller sections: Hunters Moor, Dukes Moor, Little Moor and Castle Leazes. The area is thought to have originally been covered by an oak wood. In the medieval period the trees were used for the building of houses and ships,[2] and the area was transformed into grassland for the grazing of animals belonging to artisans and traders within the town. Grazing on the Moor is still overseen by the Newcastle Freemen to this day. That usage has shaped both the landscape and the history of the space ever since.

Topography/Features

Originally, Town Moor was predominantly flat, with the land rising gently towards Cow Hill and Race Hill. The area has been used predominantly for grazing for centuries, though there is also evidence of extensive coal mining. Some of the tracks that originally crossed the Moor were converted into roads in the eighteenth century, and in some cases, these now mark the boundaries between the various moors. The route of the eighteenth-century racecourse is also still visible in the landscape as are the remains of an isolation hospital that was constructed in the nineteenth century. By that point, a large portion of the Moor was said to have been covered by whins or gorse, which some local residents would cut and bring to town to sell for firewood. One such resident, "Whin Bobby", was immortalised in Henry Perlee Parker's "Eccentric Characters of Newcastle", which now hangs in the Laing Art Gallery in Newcastle upon Tyne.[3] Today, trees line some of the roads across the Moor and there are also one or two small isolated copses, for example, at the site of the isolation hospital, but the vast majority of the area remains open grassland. The biggest alteration to the topography of the Moor in recent times has been the creation of two large artificial hills out of spoil produced during the construction of the urban motorway in the 1970s. These are now popular for sledging in winter.

DOI: 10.4324/9781003204558-3

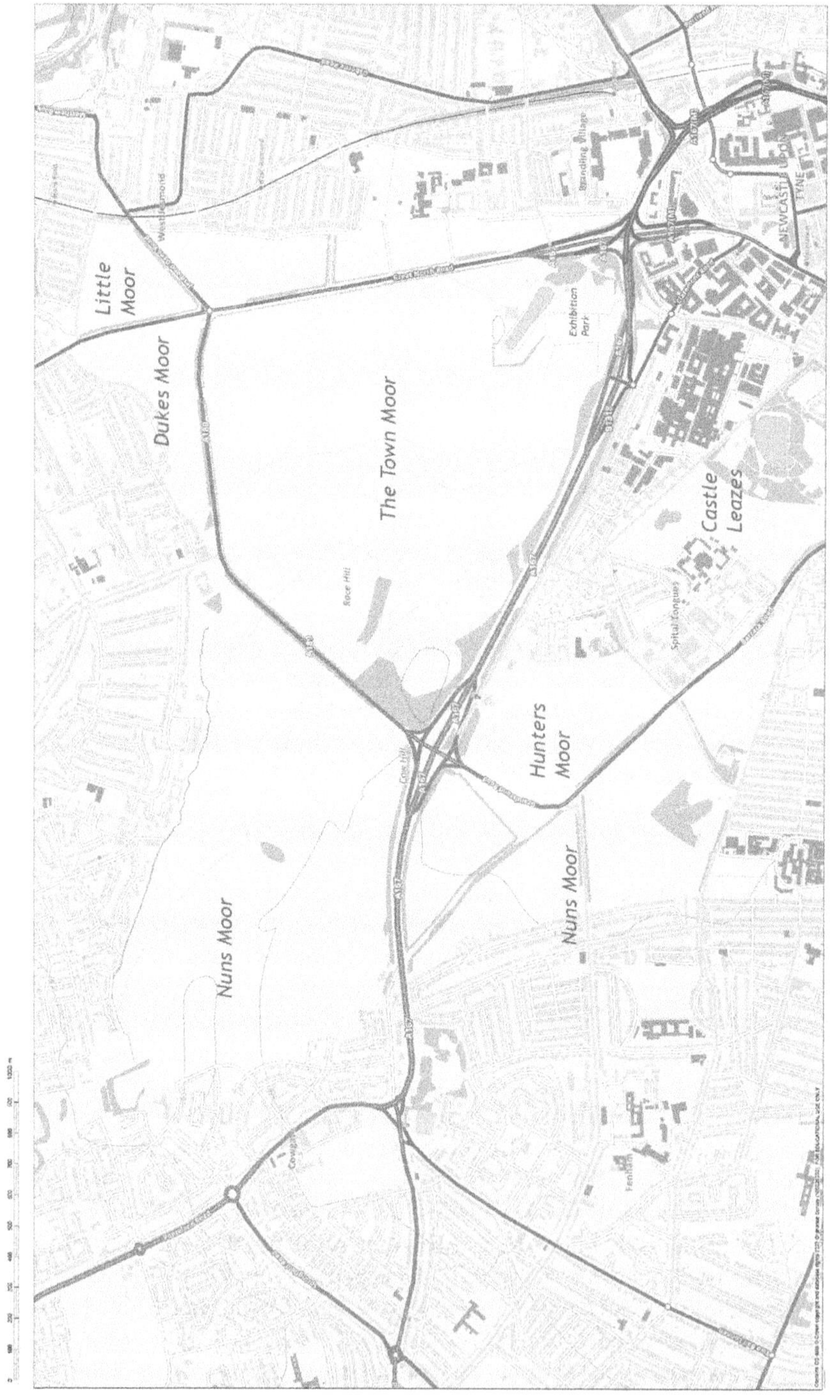

Figure 2.1 Town Moor, Newcastle upon Tyne.
Contains public sector information licensed under the Open Government Licence v3.0.

Narrative

There is landscape evidence of two prehistoric settlements in the area around Race Hill. The first may date back to the Iron Age, the second could either be late prehistoric or Romano-British.[4] From the medieval period, the area was used for both grazing and mining with evidence of early bell pit mines on lines following the coal seams as well as signs of the later use of the pillar-and-stall technique on Nuns Moor.[5] As was typical of urban commons, recreational activities became important from the eighteenth century.

That era was also marked by conflict between the Freemen and the Newcastle Corporation over the ownership and management of the Moor. Following a court case in 1773, prompted by a dispute over the enclosing of a plot of land by the Corporation, the Town Moor Act of 1774 laid down a joint ownership structure, according to which the Corporation owned the land but the Freemen controlled grazing rights. The Act also detailed the nature of those rights: individual Freemen and their widows were allowed to graze two milk cows each on the Moor. A Herbage Committee was appointed by the stewards of the guilds to make decisions about the Moor on behalf of the Freemen.[6] The Act allowed, and laid down the regulations for, intakes – temporary leases of plots of land for cultivation – with the profits providing a fund for poor Freemen and their families. The practice of leasing out plots continued well beyond 1800.

During the nineteenth century, there were also some minor encroachments around the edges of the Moor for the construction of more permanent structures, including a barracks, two parks and a hospital. The first of these was part of a nationwide defence strategy in the context of the Napoleonic Wars. The others reflected the Victorian drive for "improvement" which was embodied in the 1870 Newcastle upon Tyne Improvement Act. Despite its achievements, this agenda generated considerable opposition and controversy.

The grazing of cattle has continued to be a notable feature of the Moor into the twentieth century. This historic use sits alongside the newer focus on recreation with locals using the Moor for walking, running and cycling. It also serves as a venue for various regular events, including the Great North Run, Pride and Mela Festivals and the annual Hoppings Fair. A new Town Moor Act was passed in 1988. It confirmed the arrangement of dual control in operation since 1774. The Act states that the City Council legally owns the freehold, but the Freemen have an absolute right to herbage, so that mutual agreement is required on all matters relating to the land.[7] The Act declared that no more than 800 cows can be grazed on the Moor at any one time and also enshrined the principle that the public of Newcastle have the right of "air and exercise" on the Moor. Again following the earlier Act, the Stewards' Committee had the right to designate any plot of up to 100 acres (except in the area used for the Hoppings) as intakes, to be used for purposes other than grazing with the rental income going to the Town Moor Money Charity.[8] Ten years after the passing of the Act, the joint committee of the Freemen and the City Council that had managed affairs relating to the Moor was dissolved and replaced by the Town Moor Joint Consultative Committee.

Governance and Property Rights – Historic

The Changing Space of the Common

The area covered by the Town Moor has not changed as dramatically as some other urban commons. A basic narrative can be traced from the early modern period, which begins with expansion in the seventeenth century. The eighteenth century saw less dramatic transformation of the space, but changes were introduced through the construction of roads and the development of a system of small-scale temporary enclosure (intakes). In the nineteenth century, the practice of temporary intakes continued, but of greater significance was the loss of small plots of land, especially on Castle Leazes, for various public amenities.

Seventeenth-Century Expansion

While Town Moor proper and at least part of Castle Leazes (originally known as Castle Field) were appurtenances of the town from 1357, other areas of Newcastle's urban common were later additions. Nuns Moor, which stretches from Barras Bridge to Kenton up the western side of the Town Moor, originally belonged to the Nuns of St Bartholomew. After the dissolution of the monasteries, it was sold by John Branxholme to Robert Brandling, a Newcastle merchant, for £20.[9] In 1650, it was purchased from Robert Brandling's descendant Charles Brandling of Gateshead by the Newcastle Corporation and annexed to the Town Moor. Though part of Castle Leazes was already said to have belonged to the town "time-out-of-mind" by the reign of Edward III, land originally held by Thomas Davison in this area was transferred to the Merchant Adventurers and then sold to the Corporation in 1681. Twenty years later, another portion of the Leazes passed to the Corporation from Charles Clarke, a draper. According to a plan produced by the Corporation in 1731, Castle Leazes at that point comprised just over 127 acres. The northern boundary of the whole area is marked by a stream that runs from Kenton in the west to Coxlodge in the east. Elsewhere, the edge of the common was probably determined by parish boundaries. Boundary stones were erected after the dispute in the eighteenth century. One of these survives at the western end of Nuns Moor close to the A167,[10] others are recorded on the 1864 OS map and metal boundary plates are still visible along the wall between Grandstand Road and Coxlodge.[11]

Road Building

Tracks used by humans and livestock crossed the Moor throughout its history, but in the eighteenth century, some of these were made into more robust thoroughfares. In 1747, improvements were made to the old cow causeway that ran from Barras Bridge north across the Moor. It was part of the Great North Road that connected London with Edinburgh, but by the 1740s, it was overused and poorly maintained such that it was deemed a danger to travellers. In July 1747, the Newcastle Corporation issued an advertisement inviting bids for the construction of a carriageway "to be eleven yards in breadth, and to be ballasted in the same manner as a turnpike road".[12] The same year, Parliament approved the first turnpike for Northumberland, which would involve repairs to the road from the cow causeway to Buckton Burn near Belford. The Corporation's section was completed by 1749. Four years later, work began on the Ponteland Turnpike which ran west from Gallowgate to West Cowgate out towards the village of Ponteland.[13] Grandstand Road seems to have originated as a route for the carriages of wealthy racegoers from the Great North Road to the grandstand and enclosures. Of course, road building did not end in the eighteenth century. In the 1970s, the urban motorway was constructed along the south-west boundary of the Town Moor and dissecting Nuns Moor to Cowgate. This provoked considerable controversy among locals.

Intakes

Though roadbuilding proved controversial in the twentieth century, it appears to have been less so 200 years earlier, with the Morpeth and Ponteland routes provoking little discussion or opposition. The other eighteenth-century innovation – intakes – was by contrast more contentious. The haining (enclosing) of the Leazes "for the benefit of the herbage" was first proposed by Newcastle's Common Council in 1710.[14] The issue arose again during Sir Walter Blackett's term as Mayor of Newcastle in 1771–1772. The Freemen sent a petition to the magistrates requesting that part of the Town Moor be enclosed and let for cultivation, with the money raised being used to relieve indigent Freemen and widows. Blackett claimed that it was not within the magistrates' power to grant this, but soon after, the magistrates themselves proposed leasing out part of the Moor.[15] On 31 December 1771, the Common Council discussed a plan to lease out a plot of c.89 acres for improvement, with the revenue being distributed among poor Freemen. Adverts appeared in Newcastle newspapers from 4 January 1772, and 16 days later, Joshua Hopper, a Quaker from County Durham, was granted a lease for an 89-acre plot to the west of the Ponteland Turnpike between Gallowgate and Cowgate for 12 years on condition that he improve the land using appropriate agricultural methods. The controversy that followed will be discussed in detail in Chapter 7. It led to the court case which resulted in the 1774 Town Moor Act. Despite these controversial beginnings, the practice of renting out plots of land on the Town Moor for improvement became common practice through the late eighteenth and nineteenth centuries.

The 1774 Act set out the regulations according to which land could be leased and confirmed that the profits were to be distributed by the guilds among infirm Freemen and their widows. Certain areas, including Cow Hill and the Racecourse, were to be exempt from letting and preserved as public amenities. Intakes continued to be leased through the nineteenth century, but in the 1860s, the annual meeting at which plots were distributed became an opportunity for opposition to be voiced against wider changes and improvements, including the idea of turning part of the Moor into a "People's Park".[16]

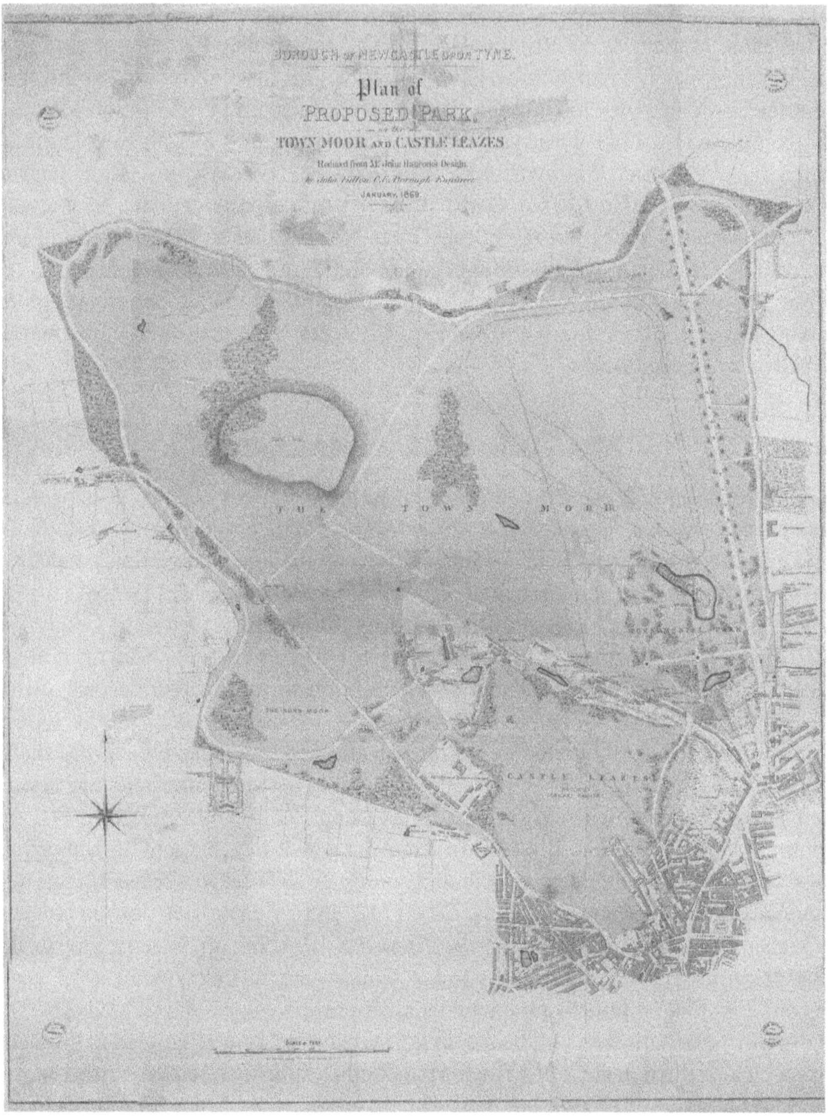

Figure 2.2 Plan of Proposed Park, Town Moor and Castle Leazes. Reduced from Mr John Hancock's Design by John Fulton, January 1869. Newcastle Libraries Collection.

Encroachment in the Nineteenth and Twentieth Centuries

Most encroachment has been small scale and limited in scope. Only a little over 200 acres have been lost since the early nineteenth century. That reduction has been the result of various encroachments made by mutual agreement between the

Freemen and the Council for purposes that were deemed to be of benefit to the city and its residents. In 1805, the Mayor and Freemen granted five acres of the Leazes (later increased to 11) to the government for use as an artillery depot. Two stable blocks were constructed together with barrack rooms, officers' residences and various other buildings. Further land was taken following the Newcastle Upon Tyne Improvement Act of 1870 to form parks and, in the 1890s, the Royal Victoria Infirmary. As its name suggests, Leazes Park, which opened in 1873, was built on part of Castle Leazes. The Improvement Act also made provision for a second park to be established on the site of the former bull park. It began to be used for recreation from the 1870s and was the setting for the exhibitions in 1887 and 1929. In 1930, the majority of the buildings that had been built for the 1929 exhibition were demolished, with the exception of The Palace of Arts; at the same time, the name of the park was changed to Exhibition Park.

Encroachments have also been made to provide sporting and academic facilities. In 1892, part of Nuns Moor was appropriated for the course of Newcastle United Golf Club (unconnected to the more famous football club). The course is still situated there today. In the 1960s, Newcastle University acquired land on Castle Leazes to build the eponymous halls of residence. This land was offered in exchange for 30 acres at Little Benton. The most ambitious plan of this kind was, however, rejected. In the 1990s, Sir John Hall proposed a multimillion pound development in Leazes Park for Newcastle United Football Club, whose ground at St James' Park abuts the Leazes, and other sports teams within the city. The plans were, however, withdrawn due to opposition from the Freemen and local residents.

Oral histories of Town Moor conducted for this project revealed that the space is an important part of Newcastle to the interviewees and has been for decades:

> I'm pretty sure you know … you'll get into trouble from the Freemen for calling it an urban common … We like to call it the Town Moor and it's unique, it's not actually a common. But you know the background to all of that, you know why we are so passionate about it, or most of us are passionate about it … We feel quite protective of it.[17]

Several interviewees remembered campaigns they staged when they were younger, including "Save our Trees, Save our Moor", a campaign of ribbons tied across the space.[18] Another group, Save Our City from Environmental Mess (SOCEM), would occupy the trees to prevent felling, and one former member argued that "the Great North Road … is still probably narrower than it might have been … had SOCEM not had some influence". [19] Threats to the Moor were and continue to be met with resistance, and one interviewee reflected that through these protests, you could see "just how precious that area was to people … I think there's, there's a feeling that, well, if we don't use it, we'll lose it".[20]

Pride in the Moor was in the interviewees often connected with pride in Newcastle as a city, with an emphasis on how special it is to have such a large green space in the centre of an urban industrial area. One interviewee underlined that to know the history of the Town Moor meant understanding the history of Newcastle "going back to King John and 1216":

> That's where the coal was ... that provided, you know, sort of the fuel for the town came from, you know, originally bell pits and then, ... sunken pits. And of course the Town Moor was also ... a wooded area originally, a lot of oak trees and they were cut down to provide for buildings in the city and for ships for the Tyne. So it's always been, there's always been a sort of a very close link with the history of Newcastle.[21]

As with our other case studies, it seems clear that a vital part of ensuring the continued preservation of common land is the investment and inclusion of the community that it serves:

> I mean, I suppose Newcastle people are proud of their city, but every citizen is proud of their city, I suppose, ... And I suppose they're proud of things that make it different, and the Town Moor makes Newcastle different.[22]

The Urban Common in Social and Political Context

Unlike many urban commons, Newcastle's Town Moor was not originally manorial waste, yet it is one of the few on which a traditional common right – the grazing of cattle – has been exercised continuously since medieval times. In 1827, there were approximately 600 cows grazing on the Moor, with the maximum allowed in any year being 800.[23] This is in line with current practice; the Town Moor Act (1988) also set the maximum for a year at 800 and approximately 600 cows were grazed there in 2018.[24] The uses to which the Town Moor has been put over the centuries relate to its distinctive origins, as an appurtenance of the town granted officially to the Freemen in 1357 and to the longstanding presence of the cows.

Common Rights and Resources

The Freemen have always managed the grazing rights. As early as 1562, a "Viewer of the Common Moor" was appointed by the Freemen to oversee grazing, and grassmen were employed to impound cattle and remove trespassers.[25] The duties of the town "neateherd" or "noltherd" and his four servants, whose role it was to tend the cattle on behalf of the Freemen, were set out in the Common Council Act Book in 1653.[26] These men would enter the town every morning, each via a specific route, blowing their horns to indicate to the Freemen that it was time to bring their cattle onto the Moor. They would tend the cattle during the day before driving them back into town in the evening to be reclaimed by their owners. Printed rules from 1823 indicate that noltherds were still employed to receive the cattle at the town gates each morning and return them in the evening, as well as having responsibility for the bulls in the bull park and for the maintenance and improvement of the land, hedges, ditches and ponds.[27] Two workers' cottages were built to house those employed by the Freemen to maintain the Moor.

One was located at the junction of Grandstand Road and Ponteland Road. It has now been demolished, with only low banks remaining where the wall around it once stood. The other survives on the corner of Dukes Moor. Its distinctive colouring gives the Blue House Roundabout its name.[28] Today, the Freemen employ four full-time agricultural workers to engage in routine maintenance and improvement.[29] Livestock fairs were also held regularly on the Town Moor. Elizabeth Halcrow claimed that a Lammas Fair was established by King John and held annually on 1 August and that from a later date Cow Hill Fair was held for the sale of livestock.[30]

Grazing was not the only traditional common right exercised by the Freemen on the Moor. Mining also took place from as far back as medieval times. Archaeologists have identified distinctive ring banks of spoil from early bell pit mines and a document dated 1562 refers to the sinking of pits.[31] In the tradition of common rights, the mines were to be worked for the benefit of the Freemen and to ensure cheap coal for local citizens, not as commercial enterprises.[32] Collieries operated on the Moor from the seventeenth to the nineteenth century.[33] At the centre of Nuns Moor are earthworks that indicate the use of the pillar-and-stall technique, and there is further evidence of this technique elsewhere on the Moor, as well as of twentieth-century opencast mining and of clay mining on Nuns Moor.[34] Mark Bowden has suggested that there is "an almost complete coal-mining history in the field archaeology of Newcastle Town Moor".[35]

The open space of the Moor was itself a resource that was put to various uses. Despite the continued history of grazing cattle, the Town Moor is typical of English urban commons in having a long history of recreational activity.

Recreation

Newcastle's racecourse moved from Killingworth Moor to the Town Moor in 1721, and thereafter horse racing was a regular feature on the Moor for more than a century and a half. [36] Races initially took place annually in the week before Whitsun (to coincide with the assizes), but from 1751, they were held at Midsummer and accompanied by an annual fair, with race week being observed as a local holiday. In 1756, the Common Council granted £75 to improve the race ground. Evidence of both the original and the newer course remains in the landscape.[37] Initially, facilities consisted mainly of marquees and temporary wooden structures, but in 1800, a permanent stone grandstand was built at the north end of the course. It was rebuilt following a fire in 1844, and remains of the building are evident in the current built environment. By the late nineteenth century, the number of people present on Race Wednesday may have been as high as 80,000–100,000.[38] In the 1880s, the races moved to Gosforth Park.

Part of the reason for the move was concern at the "immoral" behaviour and criminal activity that took place during race week. An article in the *Newcastle Journal* in July 1843 called for an end to the practice of erecting tents for the sale of alcohol on the Moor around race week. The result, the author noted, was "a scene of drunkenness and immorality".[39] Yet, the authorities were reluctant to punish large numbers of people for the actions of a few.[40] Alongside drunkenness, illegal betting also provoked concern. In July 1879, suspects were charged under the Betting Act for activities during that year's race meeting.[41] Other criminal activities resulted from the gathering of large numbers of people. In 1865 a local woman, Mary Cowans, had her locket snatched during the races and in a separate incident three Gateshead men were charged with riotous behaviour.[42] Moving the races to Gosforth was presented as a solution to these problems. As a correspondent to the *Newcastle Courant* commented in 1888, while the activities that took place at Gosforth were no less objectionable, the fact that they are less public is a distinct gain. They are a good distance from the town, and as to attend the races necessarily involves some expenditure of time and money, the stoutest opponent to the institution will admit that they are less likely to do harm than formerly.[43]

> In a bid to deter certain members of the community from venturing to Gosforth, a festival continued to be held on the Town Moor in late June to provide a sober replacement for the annual "carnival of the North".[44] Inspired by a recent visit by the temperance missionary Richard T. Booth, June 1882 saw the first Temperance Gala on the Moor, comprising sports for all ages, brass band contests, a fair and a treat of tea, sports, and games for over 1,000 poor children, as well as temperance speeches and the signing of a petition.[45] Supporters claimed that the event proved popular, with 140,000–150,000 people said to have attended in the first year and as many as 300,000 subsequently.[46] The Temperance Gala was held for many years and the Hoppings Fair, still held annually on the Moor today, continues the tradition.

Other, more modest, forms of entertainment also took place. Cricket matches and kite competitions were both popular in the nineteenth century.[47] Not all recreation was sedate and uncontroversial, however. In the mid-nineteenth century, there was much consternation over pitmen from local villages bowling on the Moor. Several accidents were reported, and in April 1854, a notice was issued prohibiting such activity.[48]

At the end of the nineteenth century, the Barnum and Bailey circus was held on the Town Moor.[49] This was also the era of exhibitions, with the Jubilee exhibition of 1887 and the Great North East Exhibition of 1929 both held in the bull park. The latter was a grand affair with the construction of a festival hall with seating for 1,400 people and three palaces celebrating science, the arts and industry. The reservoirs were connected to form a boating lake. Most of the buildings were demolished, but the Palace of Arts was allowed to remain and now houses the Wylam Brewery.

Political and Military Uses

Like other urban commons, Newcastle's Town Moor was used for political and military purposes. Political meetings took place during the nineteenth century, including a Chartist meeting in April 1848 and reform demonstrations in the 1860s and 1880s.[50]

One of the earliest recorded instances of military activity was the camp of 15,000 soldiers under General Wade during the 1745 Jacobite campaign.[51] In 1805, military activity gained a more permanent tenancy with the creation of the barracks.[52] In the late nineteenth century, military reviews and mock battles were held on the Moor.[53]

During World War One, as well as troop inspections, the Moor was used by the airforce. Armstrong-Whitworth secured contracts from the War Office to build planes and established an aircraft factory on Dukes Moor, taking over the racecourse grandstand. The Moor was also used as a holding ground for aircraft involved in the aerial defence of the north-east coast, and in 1919, the Moor hosted an exhibition of some of these planes.[54]

The space was put to more extensive use in World War Two. Aerial photographs reveal the remains of a radar station on the north side of the Town Moor[55] and of a prisoner of war camp at the eastern end of Nuns Moor.[56] Practice trenches were also built along with part of the northern stop line which comprised pill boxes and tank obstacles. Parts of the Moor were also used for allotments created as part of the "Dig for Victory" campaign.

The Darker Side of the Moor: Waste Disposal and Death

The perception that the Moor was an "empty" space led to it being used for darker purposes, including the disposal of unwanted material. Towards the end of 1867, the police seized nine cases of nitroglycerine from the beer cellar of the White Swann Inn in Newcastle's Cloth Market.[57] It was suspected that this highly explosive compound was intended for use by Irish terrorists. The authorities decided to dispose of it by dumping it in a disused pit on the Moor. The Sheriff of Newcastle, Mr John Mawson, together with the town surveyor, Mr Bryson, Constable Donald Bain and several others, carried the cases in a cart from the town onto the Moor. Soon after the liquid had been emptied into the pit there was an explosion. Mawson, Bryson and Bain were all killed along with four other men.

These were not the only deaths to take place on the Moor. A newspaper article from May 1859 names 15 people executed on the Moor between 1751 and 1844.[58] It was also, occasionally, the setting for murder or suicide. In 1894, the body of a young man whose throat had been slit was found on the Moor.[59] Later that year, newspapers across the country reported the tragic murder of 32-year-old Annie Harding.[60] Her body was found close to the bull park early on Sunday, 1 July. Like the young man, her throat had been slit and the inquest, finding evidence of a desperate struggle, returned a verdict of "Wilful murder".[61] In July 1895, a man was found lying in a shed on the Moor at 10 p.m. on a Sunday. He was roused, but subsequently collapsed and died of suspected laudanum poisoning.[62] On the morning of 8 November 1900, Alfred J. Sleight, a butcher, was found on the Moor with a self-inflicted bullet wound to his head.[63]

Contemporary Understanding of the History and Management of Newcastle Town Moor

The history of the Town Moor is directly interlinked with the history of the Freemen of Newcastle. The cows that graze from spring to autumn have become a symbol of this lasting connection, and, to some, the reason why Newcastle still has such a large urban common. Threaded throughout the oral histories for this case study is pride in the Moor and a belief that it makes Newcastle different. One Freeman reflected on this relationship:

> Well, we look on it as the lungs of the city … the Freemen have looked after the Moor for some 800 years plus, and hopefully we can look after it for the next 800 years … So, we're very proud. It's an open space for the city … I mean where else in this country of ours is there 1,000 acres bang in the middle of a town? So, we're very proud … we do it because of our love of the city and the moor.[64]

It was clear during the interviews that the Freemen are considered to be integral in the preservation of the Moor in its current iteration, regardless of whether there were concerns about them as an institution:

> … the Freemen are essentially a medieval institution that don't come out of the democratic background. They have the advantage of being so immoveable as an institution that they probably prevented some Labour politicians in the past like T Dan Smith from building on the Town Moor. So, from that kind of, again with a small 'C', conservatism they almost certainly have to be credited with saving the Town Moor. I don't think anyone could take that away from them, really. But they are a very, from the outside, they are a very closed organisation.[65]

This perspective was common among our interviewees. One referred to them as the "main anchor in the survival of the Moor".[66] Another interviewee reflected that the "Freemen haven't got much power other than just retaining, with the city … just grazing rights really".[67] Of utmost importance was the link between the herbage rights and the continued preservation of such a large urban space. As one interviewee said, "I think the way to understand the Freemen is to think of cows. It's all about cows and cattle".[68]

One member of the Freemen commented that he felt "quite strongly" that the grazing rights were important for the preservation of the Moor. He emphasised that Freemen "don't own the Town Moor" but that "we own the grass, so we have the right to graze on the Town Moor … It's such an important asset."[69] The grazing rights are often seen as increasingly symbolic, as Freemen do not graze their own cows, but instead often lease their stints out to farmers. One Freeman explained that in order to "look ahead", they've "got to keep cattle on the Moor for grazing". And for that reason, the Stewards committee purchased 50 head of cattle, because "the one thing we have to do is keep the cattle grazing on the moors".[70] This underlined the perceived importance of the history of grazing to the protection and preservation of the Town Moor. Without the cows, many believed, there would be no common.

With regard to the future of the Moor, one Freeman voiced her concerns about the relevance of the beef herd in a society that is eating less meat. She argued that beef grazing was "increasingly unpopular" and that one of the new challenges for the Freemen would be recognising the new "trend towards vegetarianism and veganism and rethinking the strategy for existing".[71] Another concern was the lack of young people taking up their position in the Freemen or their guild. Eligible candidates can join in their 21st year, but as one freeman commented:

> how do you keep a twenty-year-old bothered about making sure that we comply with a 1988 Act that means we've got grazing rights on the Town Moor. How do we make that attractive to a 20-year-old and how do we make those 20-year-olds come to guild meetings and come to company meetings?'[72]

One Freeman argued that the organisation needed to "speak the same language" as the younger members and recognise some areas where tradition could be challenged. For example, an interviewee cited a rule that prevents "children born out of wedlock" from becoming Freemen, which (it was felt) reflects an outdated perspective on marriage. As marriage increasingly becomes an "outdated institution", perhaps rules like this prevented the Freemen from attracting younger members.[73]

Awareness of the structure and management strategy of the Freemen varied from interview to interview. When asked "who owns the Moor?" the answers varied. One community organiser stated:

> Ah, now that's interesting. It's very interesting because I believe it is owned by the … or at least the guardians of it are the Freemen of Newcastle, which are this weird, medieval group. So I think that they own it, but it's very complicated in Newcastle because I don't know if they hold it on behalf of the people of Newcastle … I know anyone is allowed to graze their cows on it … So I don't know if they hold it in trust for the people of Newcastle or they physically own it and have … a certificate of freehold over it.[74]

Another stated that "it seems all kind of in the mists of time … but I don't really understand the complexity of that". She clarified that from her experience of living in the city, "the Freemen seemed to have had more power than the council to make changes or not make changes".[75] A third remembered being told from a young age that the reason it hasn't been built on it was "because it was protected by the Queen", but was unsure whether that was a myth.[76] The interviews for this project suggest that the lack of publicity for the Moor's management plan and closed meetings have perpetuated uncertainty about the role of the Freemen and the status of the Town Moor.

Some interviewees reflected critically on the role of the Freemen and the future, asking whether a more democratic approach should be adopted to reflect modern Newcastle:

> So I think that demand's falling off and eventually the Freemen will be overtaken by the circumstances. There'll be fewer and fewer cattle, I think. They'll have less … they'll have fewer grounds to keep closing things like allotments. They would realise that the world has changed, that the residents are much more aware of the environment or issues and concerns in the world. And grazing cattle on a couple of a thousand acres of green grass is not necessarily the best way of meeting those.[77]

This interviewee called for a new Town Moor Act that would rebalance "the power between the people as represented by the council and the Freemen and their 2000 people". He clarified by stating:

> People tend to think of the Freemen as … as a really good thing … And it is great that we've got this thousand acres at the centre of the city that isn't built up or commercialised, and that's fine. I don't want to challenge that at all actually. But there's room there for other things … We should have more biodiversity, we should be growing our own food. We should be encouraging people to do these things. It's good for the health, and it's good for the mental health as well. And we should be doing more of these things and not less.[78]

Another interviewee commented on the disparity between the Labour Council and the structure of the Freemen, arguing that the "vast bulk of the city is public green space in the hands of two, effectively two private, self-appointed oligarchies … I think [it] is historic, really".[79]

Contemporary Governance and Property Rights

The contemporary governance of the Town Moor is shaped and regulated by the provisions of the Newcastle upon Tyne Town Moor Act 1988. As has already been noted, the 1988 Act replaced the 1774 Act, with the express objectives, *inter alia*, to: modernise the management and administration of the Town Moor; in relation to public access and recreation to enact provisions "for and in relation to the enjoyment and improvement of the surface of the Town Moor"; and "to make provision for the better organisation and regulation of the freemen" of Newcastle.[80] The freehold and soil of Town Moor remain vested in the City Council, subject to the herbage rights of the Freemen of Newcastle.[81] The presence of beef cattle grazing on Town Moor is one of its unique and defining features, and the Freemen's herbage rights over the Moor are central to the way in which its management is structured under the terms of the 1988 Act.

Joint Management and the 1988 Act

The Act establishes a "joint" management model for the Town Moor, under which rights and responsibilities are shared between the City Council and Freemen in a nuanced balance. The management of the Moor is entrusted by the 1988 Act to the Stewards Committee of the Freemen. The Act provides for a committee of stewards and wardens to be elected by the Freemen each year of not less than nine and not more than 12 of their members. The Stewards Committee has the duty to act on behalf of the stewards, resident Freemen and widows for all purposes in relation to Town Moor and its maintenance.[82] This is a wide-ranging supervisory and management remit, as the 1988 Act defines "maintenance" for these purposes to mean:

> the care and supervision of top soil and of the exercise of herbage rights over the Town Moor and the husbandry, draining, fencing and cleaning of and the

> carrying on of agriculture on the Town Moor for the purposes of grassland management and the employment of persons for those purposes.[83]

The City Council can construct new roads and paths across the Moor, but only with the concurrence of the Stewards Committee. And all roads and footpaths across the Moor remain free of the Freemen's herbage rights and are maintained by the City Council, as are the existing trees on the Moor.[84]

The Moor provides a large area of recreational urban green space in the centre of Newcastle. Its importance in this regard is underlined in the Act, which provides that it must be maintained as an area of open space in the interests of the inhabitants of the city. Here again, however, the balance between the rights of the Freemen and those of the wider public is evident, in that its function as a protected open space is expressly "so that it shall continue both to satisfy the herbage right and to afford air and exercise for the enjoyment of the public".[85] The way in which the recreational rights of the public and the exercise of grazing rights are balanced is central to the role and utility of the Town Moor as a "green lung" for the city. It also shapes the community's sense of "ownership" of the Moor as a public space in the heart of the city, contributes to its sense of place belonging and identity, and dictates public attitudes to – and use of – the Moor for different recreational purposes. These were explored further in the oral history research conducted with recreational users and user groups, discussed further below.

The Stewards Committee has responsibility for regulating the grazing on the Moor. This is organised around a system of annual "stints", each stint giving the right to depasture one cow on the common. The maximum number of cattle that can be depastured on the Moor in any given year is determined by the Stewards Committee, but cannot exceed 800.[86] The stints that are available each year are allocated to Freemen and widows on application. Each resident Freeman and widow is entitled to apply for stints before 1 April annually; if the number of applications exceeds the number of stints available, then a ballot is held by the Stewards Committee. The allocation of stints is to be made in equal numbers to those Freemen and widows who apply, insofar as the number of stints available allows for equal distribution.[87] The grazing rights conferred by the allocation of stints can be exercised by a Freeman or widow grazing their own cattle or transferring the stint to a third party such as a local farmer. They can also authorise the Stewards Committee to sell the stints on their behalf.[88]

In all other respects, the management of the Moor is carefully structured by the 1988 Act around the joint management principle, allocating and balancing the joint responsibilities to the Freemen and City Council. So, for example, the Act sets out an "intake" system whereby land can be taken from the Moor and leased for specific purposes. The maximum that can be subject to intake at any time is 40.5 hectares.[89] Areas of intake are designated by the Stewards Committee, and these are then leased by the City Council (with the concurrence of the Stewards Committee) for periods not exceeding 21 years following a public auction.[90] The City Council also enjoys a number of important powers and functions relating to the management of the Moor, but these are in most cases only exercisable with the concurrence of the Stewards Committee of the Freemen. So, for example, the City Council can (with the concurrence of the Stewards Committee) enact bye-laws to regulate a wide range of activities and uses on the Moor, control of undesirable land uses, anti-social behaviour and a wide range of other user rights.[91] These powers have been used to introduce bye-laws placing restrictions on motor vehicles and bicycles and prohibiting their use on the Moor otherwise than on public paths and roadways.[92] Bye-laws can also protect the herbage rights of the Freemen and regulate, prohibit or restrict the use of land for purposes other than the exercise of the herbage rights conferred on the Freemen. The City Council can also, with the concurrence of the Stewards Committee, enclose or use parts of the Moor for a variety of purposes,[93] including exhibitions, recreational activities, sports or entertainments, permit the erection of buildings or structures on the Moor or permit parts of it to be used for allotments or recreational facilities.[94] These powers have been widely used to create allotments and to hold a variety of recreational and entertainment events such as the annual Hoppings fair held in June each year.

Environmental Governance

The tradition – and ongoing contemporary practice – of grazing cattle on the Moor also shapes and determines its ecology and the manner of its environmental governance. Town Moor does not host any protected areas of national or international importance or significance (e.g. any Sites of Special Scientific Interest [SSSIs]), but it is a designated Site of Local Conservation Interest.[95] Its conservation and wildlife value was assessed in a survey made in 2013 for the purposes of the Newcastle City Development Allocation Plan. This concluded that while it provided considerable value to the local community and was well used, it was "valued for wildlife but with limited management especially for wildlife".[96] Management has historically been focussed on maintaining a viable grazing regime, and the importance of what is essentially a primarily agricultural land use determines the biodiversity and ecology of the Moor. The 2013 report on the value of the Moor as a site of local conservation interest drew attention to the use of artificial fertilisers, herbicides and drainage which have all significantly reduced the nature conservation value of the Moor in recent years.[97] This style of management has created and perpetuated perhaps the most unique of Town Moor's characteristics – creating a rural character in a highly urbanised context. But it also means that its biodiversity is characteristic of an agricultural landscape, where the style of management limits species diversity, and not that of a landscape actively managed to encourage greater biodiversity.

The Moor's active management for grazing does, however, provide other significant environmental benefits. Town Moor is, for example, a key element in flood risk mitigation for the city.[98] The different sectors of the Moor all have drainage provided and maintained (at, in some cases, considerable expense) and underground water storage tanks are to be found in Fenham and under Grandstand Road. A number of initiatives to diversify the Moor's wildlife value were suggested in the review for the 2013 development allocation plan exercise. These included the extension of woodland plantation areas on the Moor, the enclosure of grazing cattle and the reduction of fertilisers on some sections of the Moor.[99] The Freemen have a strategic plan for management, with a medium-/long-term planning focus of 50 years plus, and a number of these options are being actively assessed.[100]

Contemporary Use and Recreation

The Town Moor's role as grazing land generated frustration amongst some of the interviewees for this case study. An example of the friction that this can cause was the destruction of a plot of allotments on the Moor in 2015.[101] Several interviewees commented on the Freemen's ability to remove functioning allotments without what they regarded as adequate discussion or warning:

> There was no consultation with residents, none whatsoever. They just went ahead and did it.[102]

The destruction of these allotments was considered by some to have been conducted in an inappropriate manner. The Freemen's alleged view that the allotments were disheveled, dangerous and a hotspot for crime was commented upon by several interviewees. One interviewee claimed that the Freemen were great lovers of "neat and tidy" and that the ramshackle nature of the allotments, with structures and sheds, didn't fit the concept of model allotments, where all the plots are laid out neatly, aren't built up and have a clear line of sight.[103]

Another interviewee reflected that the allotments were disorganised, but that:

> there were a lot of trees there and suddenly one day the trees were chopped down, the allotments ..... and all the structures on them were being flattened.[104]

She reflected that there were rumours that the allotments were going to be restored, but that "one suspects that it's going to be more grazing land". Great concern was caused, she claimed, by "the number of trees that were cut down … without any discussion with local people as far as I could make out".[105] A local councillor reflected that the dispute was a "classic allotment war" and that "some people fell out with each other, gnawed away for years" but argued that "allotments classically are pretty higgledy-piggledy … It's almost the beauty of allotments".[106] The controversy around the allotments and the removal of these plots also demonstrates that the sense of communal ownership of common space remains strong. The desire to grow food, to garden and to be outdoors without restrictive policies emphasises the importance of urban common land, especially in locations where residents may not have access to land elsewhere. The frustration over the percieved lack of transparency also reflects how invested local residents are in the Moor and how much they value the concept of common space.

Another concern that appeared frequently in these interviews was the perception of Town Moor as a monoculture. Some of the Freemen interviewed reflected upon this:

> The other thing is certainly around biodiversity. We need to make sure that whilst the Town Moor has a specific purpose and it's enshrined in law, and we would always want to keep it as grazing land, we have to make sure that we fit in with the rest of the world and plant some trees, plant some wildflowers and make sure we look after the wildlife. There's quite an abundance of wildlife. We've got a little wooded area and stuff like that. So that all needs to be properly managed and it needs, we need to make sure we're doing our bit on that front … Particularly these days when people are starting to wake up to biodiversity and what not. It's an important asset to have, and I think Newcastle are extremely lucky to have this massive asset.[107]

One interviewee commented that even if there were no cattle, the Freemen would still own the herbage of the Moor, and therefore the right to do anything, even rewild. He concluded that "they don't want to, but they could".[108] Again, this points to the strong connection between the grazing rights and the preservation of the Moor. Rewilding to some could erase the protection ensured by the Freemen's right to graze.

A local community group has been monitoring the wildlife in their area of the Moor and were developing ways of adapting to an urban environment:

> …one thing we've visually logged in our minds now is in watching the birds feeding off the insects, is how important the undergrowth is for producing the insects, supporting the insects that the birds feed off. And that's already completely changed the way we were thinking of managing the community orchard. We planted it in a remote part of Nuns Moor Park, because when we planted stuff elsewhere it got damaged or stolen. So, we hit on the idea of a stealth orchard, so an orchard that's hidden, weed grass. And it survived really well, a whole collection of fruit.[109]

Several interviewees raised the subject of rewilding, but more frequently interviewees discussed varying the wildlife on the Moor so that the grazing could co-exist with a range of other habitats. One interviewee claimed that rewilding was "a lovely idea, but [was concerned by] the amount of crime that could bring about…"[110] and emphasised that the grazing was an important aspect of the Moor. But he also commented that it was "a biological desert".[111]

The Town Moor has little in the way of infrastructure, but this is itself a factor in its popularity. It is a space that is flexible, available and free. Interviewees commented on its convenience for sustainable travel, its role in tackling mental health, its function as a wildlife corridor and the community of the Parkrun. The importance of the Moor seems to lie with its ability to be anything, for anyone. One interviewee reflected on the new idea of social prescribing and outdoor space for people's mental health concerns:

> as a GP I worked in the city mostly and I used to really worry about people … [and] think … how can they be healthy, and mentally healthy in particular, when they never have anything other than this … and the area I was in was not green at all and … there's now evidence coming through in research that, that, you know, just having greenery, helps people's wellbeing, mental well-being, and having, you know, more than just … a kind of wizened tree … you know, actual space … We evolved from creatures that had space around them and greenery around them and interacted with it all the time. That's how they made their living. And it's only in the last, I don't know, a couple of hundred years probably that we haven't, most of us haven't had that.[112]

The Moor plays host to an increasing number of festivals and gatherings, including Pride and the Mela, in addition to The Hoppings, the largest funfair in Europe. Yet the majority of the interviewees spoke about their own use of the Moor for running, walking, travel and enjoying nature rather than the events themselves. One interviewee spoke about a cricket club who use a cricket pitch on the Moor which has been largely unused. He claimed that in the course of creating the club, they:

> discovered that there are all sorts of other sporting groups using that bit of the Moor from all sorts of places … groups of blokes who live in some cases in the East End of Newcastle who come over regularly to have football matches with each other. Nobody seems to collide with each other over it.[113]

This highlights the importance of educating people about public, common space and how it can be useful for well-being and community. One interviewee, when asked if the Moor needed development, replied, "No, it just needs to exist".[114]

Present Voices, Future Directions

Scrutiny of the balance between grazing rights and recreational rights on the Town Moor surfaced again in participatory research exploring users' visions for the future. Participatory design-based research carried out for this project encompassed an architectural live project developed with the Newcastle University School of Architecture, Planning and Landscape (APL).[115] Architectural live projects typically involve students moving out of the studio to engage with real-world settings. In this case, the research team collaborated with tutors at Newcastle University APL to involve students in the design of participatory engagements for public users of the Moor, which served to circumnavigate COVID-19 pandemic lockdowns. Students used a series of cultural probes, for example, postcards, design games, makeshift maps, drawing material and photographic tasks, to create participatory packs that enabled community users to chart their experiences of the Moor and reflect on what they most valued for the future. Analysis of participants' observations informed the design and installation of small, makeshift interventions on the Moor, which tested out ideas and engaged passers-by in further dialogue.

The visions for the future of the Town Moor that emerged through these participatory engagements echoed initiatives raised in the 2013 development plan,[116] including adjustments to accommodate citizens and free-roaming cattle and the management of the land to increase biodiversity. Attitudes to the co-habitation of grazing cattle and citizens pursuing outdoor leisure were vexed by feelings of anxiety at coming into close proximity with the cows. One participant expressed feeling "a little bit uneasy if near them", explaining how the cows can be "quite unpredictable if you're running or cycling".[117] Another participant identified conflicted emotions around their Moor visits: "I love getting nice fresh air, being in nature, but sometimes I get quite nervous around cows".[118] For this participant, there was a desire to have "a place to relax and not have worries you get attacks by cows".[119] Concern over safety and constraints on relaxation posed by internalised fear of attack were voiced most acutely in relation to children by parents and grandparents.

A design solution in the form of small, sheltered seating areas as an alternative to the enclosure of cattle was tested out in a temporary installation on the Moor, and received positive responses from passers-by. One contributor commented:

> Shelter would make my wife feel definitely more safe. She had a bad experience with cows and a shelter would make her feel less nervous. Great idea in case cows run towards you. Great idea to have time to get used to cows.[120]

Figure 2.3 Emma Itu. *Sheltered Seating.* Student Participatory Intervention, Town Moor, Newcastle, 2021.

Image: Diana Mihailova.

His reflection identified the capacity of a shelter to provide temporary refuge and to enable a gradual familiarisation to being with cows, a sentiment echoed by other participants, again with particular reference to children. Beyond fostering a sense of safety, participants felt the provision of sheltered seating would enhance comfort and access more generally; it would "bring more people to the Town Moor"[121] and encourage them to "feel calm and slow down"[122] to "sit and enjoy"[123] and to "prolong their stay".[124] It was clear participants' comments emphasised notions of dwelling, encouraged by both the sheltering and seating features that contrasted to perceptions of a current lack of seating. Where present usage accents mobility, either for exercise or "for scenic transit to other places",[125] there seemed to be a desire to increase opportunities to linger and gather in small social groupings, for family picnics, for instance. Dwelling, however, as one participant noted, can include "strange people hanging around".[126] So, where participants valued the shelters as an invitation to increase connection with the Moor and each other, "where they are located" would need careful consideration.[127]

The desire to extend civic place-making in the Moor beyond its role as a "transition thoroughfare"[128] was also reflected in suggestions to increase public engagement with the cultural heritage of the site. Historical anecdotes and accounts contributed in the participatory packs reflected the historical value of the Moor in the public imagination and led to ideas on how to render historical narratives visible within the landscape. Suggested interventions ranged from increased signage to inform users of significant historical events to public art memorials to re-contextualise "difficult" histories, a steel, witch-shaped structure bearing the names of the people executed for witchcraft here.[129] Though the form of intervention may vary and could include less tangible, more digital, mediums, there was an appetite to advance opportunities to learn about the Moor's history within the site itself.

Concerns over personal safety within the Moor were not limited to the fear of cows grazing freely. Participants also expressed disquiet over individual routes of access, crossing Grandstand Road at the bottom of Cow Hill, for example, was described as "dangerous", "stressful" and provoking a sense of "dread".[130] Even more significant in participants' reflections were feelings of anxiety associated with walking or cycling in such an expansive space at night. In response to a temporary light installation on the Moor at twilight that aimed to survey views on the topic from passers-by, one female contributor explained:

> I love the Town Moor but unless I'm with someone like I am now, I'll admit that I tend to avoid it at night… In the winter, it becomes a bit problematic for me when I have to make my way home around five or six pm, as it gets dark so early. That's when I tend to cycle less and use my car instead.[131]

It is clear how she modified her use of the Moor according to time of day and season to avoid darkness, significantly including a shift away from a preferred sustainable mode of transport in her commute to work. In the survey, there was a distinction between the experiences of female and male participants; the latter were more likely to "feel fairly safe walking through the Town Moor at night".[132]

However, both men and women participants agreed that an increase in the quantity and quality of outdoor lighting would improve safety.

> Even though the main parts of the Moor are pretty well lit, I wish the council invested in lighting up the areas around. I always worry about someone creeping out from the dark.[133]

The parts identified by this female contributor refer to the two main paths that intersect the Moor. Where the paths serve as popular routes of transit between the north and centre of the city, the openness of the common meant individuals felt exposed. Artificial lighting along these pathways is activated by full darkness rather than dusk, while light quality is fairly dim, dropping off sharply between path and surrounding field.[134] But where participants' views of outdoor lighting focused on perceptions of safety, the design here appears to minimise the hidden effects of light pollution on wildlife. The activation of lighting with natural darkness, use of the lowest intensity of lighting appropriate to the task and avoidance of light spill are all measures promoted as best design practice to benefit biodiversity and save energy.[135] Where this points back to the complex balance between citizens' use and biodiversity conservation, it does not suggest that issues of safety should be discounted, for it raises critical questions about equality of access to urban green spaces with regard to gender. None of the contributors reported any knowledge of violence on the Town Moor, yet for female participants, their perception of safety, or rather a lack of it, deterred them from walking on the Moor at night. According to a survey carried out by the All-Party Parliamentary Group for UN Women in 2021, however, "71% of women of all ages in the UK have experienced some

Figure 2.4 Diana Mihailova. *Bird Boxes.* Student Participatory Intervention, Town Moor, Newcastle. 2021.

Image: Thomas Coutanche, Abin John, Jake Merx, Thomas Paramor.

form of sexual harassment in a public space".[136] So where participants may not recall specific incidents, their concerns are apposite. The report encourages local decision-makers, in partnership with grassroots campaigns, to draw up plans to "work on changing wider attitudes towards what is acceptable behaviour" and to improve data collection on offences.[137]

Another aspiration that arose frequently in participants' visions for the future of the Town Moor was the ambition to diversify wildlife habitats. In a "postcard to the future" intervention aimed to survey users' opinions, comments on biodiversity and wildlife were more than double that of any other theme.[138] Wildlife-friendly initiatives proposed by participants included a "flower meadow",[139] "pond",[140] wood pile and community garden. But overwhelming opinions reflected a desire to "increase the number of trees not just on the boundary but in the middle of the Town Moor".[141] This emphasis was in part attributed to individuals' positive encounters with birdlife on the Moor, which fuelled a desire to increase bird populations and species diversity by extending woodland plantation and providing public education on issues such as the protection of ground nesting birds.[142] Calls for "more trees"[143] were also ascribed to broader understandings of the role tree coverage plays to improve air quality in urban settings and to ameliorate climate change through carbon capture. The value of woodland plantation, however, was not solely bound to concerns for wildlife conservation. Benefits to user experience were also highlighted. One participant claimed that extending tree plantation around the periphery "would help to block out some of the traffic noise".[144] Another participant commented on the play opportunities trees afford to children but concluded that present "only half alive" specimens in the copse at the bottom of Cow Hill gave rise to caution.[145]

Several participants made overtures for community orchards, emphasising their conjoined value for wildlife and citizens as habitat and mode of food production.[146] Aspirations to increase food self-sufficiency in the city were also made manifest in calls for more allotments in its largest green space. Participant observations on the lengthy waiting lists for citizens to secure a plot further illustrated demand.[147] It is indicative of a noticeable national increase in demand for allotment plots as people seek to grow their own food to ease the cost-of-living crisis.[148] One participant, who currently leases a plot from the Freemen, raised the subject of current regulations, noting "that no structure, other than for growing plants on – like runner beans, etc. – can be erected over three feet in height. No sheds or greenhouses for them".[149] The restrictions on built structures, he concluded, engendered a "unique" aesthetic, "like one big garden", but were also "discouraging (to) ambitious gardeners who want to grow tender plants in greenhouses, or store materials in sheds".[150] Again, this points to a preference for aesthetics over utility in regulations, which limit vegetable and fruit growing capacity and potentially the demographic of citizens who can make use of a plot: individuals able to store and transport tools.

The attention paid to woodlands, orchards, community gardens and allotments in participants' visions of a future Town Moor illustrates the importance placed on managing the land for wildlife biodiversity and food self-sufficiency. In harmony with the oral histories, however, participants tended to prefer the co-existence of such initiatives with the continuance of grazing cattle. But just as changing

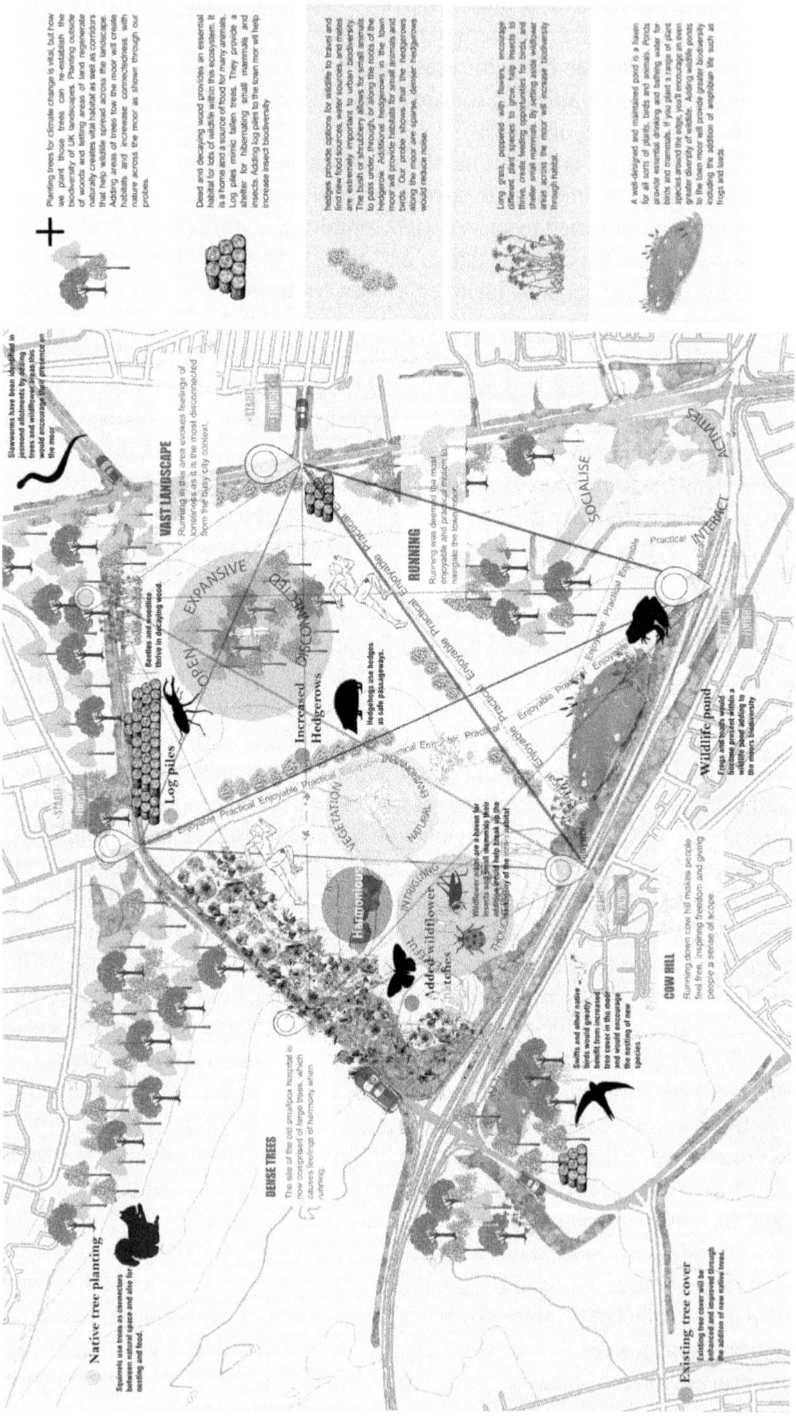

Figure 2.5 Student Group 4. *Residents of the Moor: Habitats, Hobbies and Aura* generated from responses to Participatory Pack Group 4.

attitudes to meat consumption may question the relevance of the beef herd, so do other societal concerns – climate change, biodiversity loss and most recently, food security – will exert pressure on policymakers and land managers. Where the friction between grazing rights and recreational rights remains unresolved, improving citizens' engagement with the Town Moor through education, biodiversity and community-growing initiatives may go some way towards answering citizens' calls for forward-facing solutions to the current pressing issues. The benefit of the research team's engagement with students was the capacity to see such concerns through the eyes of young people and to test out simple but potentially effective design solutions, the sheltered seating a case in point. In the long term, it is yet to be seen how land managers, such as the Freemen and the City Council, will deal with planning for future policy and climate uncertainty. But perhaps, resource might be found in both the resilience of the Town Moor and its common rights, alongside the innovation of citizens.

Notes

1 *The Freemen of Newcastle upon Tyne and the Town Moor*, Plan by P. Ansell.
2 Henry Bourne, *The History of Newcastle Upon Tyne: or, the Ancient and Present State of that Town* (Newcastle upon Tyne: John White, 1736), p. 147; and Eneas Mackenzie, *Historical Account of Newcastle-Upon-Tyne Including the Borough of Gateshead* (Newcastle upon Tyne: Printed for Mackenzie and Dent, 1827), pp. 709–714.
3 *Newcastle Courant*, 22 September 1876.
4 Mark Bowden, Graham Brown and Nicky Smith, *An Archaeology of Town Commons in England 'A Very Fair Field Indeed'* (Swindon: English Heritage, 2009), p. 14.
5 See the report by Alistair Bowden and Ian Jackson on a survey carried out on 27 June 2019, which is available at https://www.newcastle.gov.uk/sites/default/files/2019-07/Town%20Moor%20Local%20Geologcal%20Site.pdf
6 Town Moor Act 1774, see also TWAS GU/TH/16/1 Letter from Thomas Aubone to W. Brunton, dated 29 July 1776, confirming that the Act of 1774 gave to the resident Freemen and their widows their 'full right and benefit to the herbage of the Town Moor'.
7 1774 Act; *The Freemen of Newcastle upon Tyne and the Town Moor*, p. 10.
8 *The Freemen of Newcastle upon Tyne and the Town Moor*, p. 13.
9 Mackenzie, *Historical Account of Newcastle-Upon-Tyne*, pp. 709–714.
10 https://historicengland.org.uk/listing/the-list/list-entry/1106311
11 Colin Lofthouse, 'Royal Commission on the Historical Monuments of England Town Moor, Newcastle Upon Tyne Archaeological Survey Report (1995/19)', p. 9.
12 *Newcastle Courant*, 18 July 1747.
13 Mackenzie, *Historical Account of Newcastle-Upon-Tyne*, p. 709.
14 See John Brand, *The History and Antiquities of the Town and County of Newcastle upon Tyne* (London, 1789) 439.
15 (James Murray), *The Contest, Being an Account of the Matter in Dispute Between the Magistrates and Burgesses, And an Examination of the Merit and Conduct of the Candidates in the present Election for Newcastle upon Tyne* (Newcastle, 1774), pp. 24–25.
16 *Newcastle Journal*, 5 September 1861, *Newcastle Courant*, 6 September 1861, *Newcastle Guardian and Tyne Mercury*, 7 September 1861. This was also raised again in 1868 – *Newcastle Guardian and Tyne Mercury*, 5 September 1868.
17 Oral History Interview (Newcastle) 1 ("OHI Newcastle").
18 OHI Newcastle 2.
19 OHI Newcastle 3.

20 OHI Newcastle 2.
21 OHI Newcastle 4.
22 OHI Newcastle 5.
23 Mackenzie, *Historical Account of Newcastle-Upon-Tyne*, pp. 709–714.
24 'A History of the Town Moor', *Collaborate Magazine*.
25 Elizabeth M. Halcrow, 'The Town Moor of Newcastle Upon Tyne', *Archaeologia Aeliana* 31 (1953): 153.
26 TWAS: MD/NC/1/2 'Common Council Book 1636-1656', 1 April 1653, 214–215. See also MD/NC/2/2 'Common Council Book 1650–1655', 1 April 1653, 122.
27 Mackenzie, *Historical Account of Newcastle-Upon-Tyne*, pp. 709–714.
28 Bowden, Brown and Smith, *An Archaeology of Town Commons*, p. 26.
29 Ibid., p. 25.
30 Halcrow, 'The Town Moor of Newcastle Upon Tyne', p. 155.
31 Lofthouse, 'Archaeological Survey Report', p. 22.
32 TWAS: MD/NC/2/1 'Common Council Book', 11 November 1646, 72 and 73. See also Halcrow, 'The Town Moor of Newcastle Upon Tyne', p. 151.
33 Mackenzie, *Historical Account of Newcastle-Upon-Tyne*, pp. 709–714; Lofthouse, 'Archaeological Survey Report', p. 25; GU/TH/16/3.
34 Lofthouse, 'Archaeological Survey Report', pp. 22–30. See also the report by Alistair Bowden and Ian Jackson on a survey carried out on 27 June 2019, which is available at https://www.newcastle.gov.uk/sites/default/files/2019-07/Town%20Moor%20Local%20Geologcal%20Site.pdf
35 Bowden, Brown and Smith, *An Archaeology of Town Commons*, p. 37.
36 Halcrow, 'The Town Moor of Newcastle Upon Tyne', pp. 154–155.
37 Bowden, Brown and Smith, *An Archaeology of Town Commons*, p. 71. Lofthouse, 'Archaeological Survey Report', p. 12.
38 *Morpeth Herald*, 25 July 1868.
39 'The Tents on the Town Moor', *Newcastle Journal*, 1 July 1843.
40 *Newcastle Guardian and Tyne Mercury*, 21 June 1851.
41 *Morpeth Herald*, 2 August 1879.
42 *Newcastle Journal*, 30 June 1865.
43 *Newcastle Courant*, 22 June 1888.
44 *Alnwick Mercury*, 25 June 1881.
45 *Newcastle Courant*, 30 June 1882.
46 *Newcastle Courant*, 30 June 1882, *Shields Daily Gazette*, 3 July 1884.
47 *Newcastle Guardian and Tyne Mercury*, 9 July 1859 and 22 September 1866.
48 *Morpeth Herald*, 22 January 1881.
49 *Morpeth Herald*, 15 April 1899.
50 *Newcastle Guardian and Tyne Mercury*, 22 April 1848, TWAS: DF. COW/C/1743.
51 Bowden, Brown and Smith, *An Archaeology of Town Commons*, p. 50. *The Journal of the Reverend John Wesley* (New York, 1855), Volume 1, pp. 351–355.
52 Mackenzie, *Historical Account of Newcastle-Upon-Tyne*, pp. 709–714.
53 *Newcastle Courant*, 24 July 1868, *Morpeth Herald*, 25 July 1868, *Newcastle Guardian and Tyne Mercury*, 11 June 1870, *Alnwick Mercury*, 13 August 1870, *Alnwick Mercury*, 15 June 1878.
54 *Western Times*, 4 May 1917, *Hull Daily Mail*, 7 February 1919.
55 NMR: NZ 26 NW 49.
56 NMR: NZ 26 NW 31.
57 *Sheffield Independent*, 18 December 1867, *Royal Cornwall Gazette*, 19 December 1867; *Exeter and Plymouth Gazette*, 20 December 1867. The story was also covered in a number of local papers: *Newcastle Guardian and Tyne Mercury*, 21 December 1867; *Newcastle Journal*, 26 December 1867; *Newcastle Courant*, 27 December 1867.

58 *Newcastle Journal*, 28 May 1859. On executions on urban commons see: Sarah Collins, 'A Dangerously Empty Space: Crime and Punishment on English Wastelands', *Landscapes* 23:1 (2022): 48–66.
59 *Shields Daily Gazette*, 30 April 1894.
60 *Morpeth Herald*, 7 July 1894; *Manchester Courier and Lancashire General Advertiser*, 2 July 1894; *Huddersfield Chronicle; Western Times; Portsmouth Event News*; *Hartlepool Mail*, 4 July 1894; *Nottinghamshire Guardian*, 7 July 1894.
61 *Nottinghamshire Guardian*, 7 July 1894.
62 *Shields Daily Gazette*, 10 July 1895.
63 *Shields Daily Gazette*, 8 November 1900.
64 OHI Newcastle 9.
65 OHI Newcastle 6.
66 OHI Newcastle 7.
67 OHI Newcastle 4.
68 OHI Newcastle 6.
69 OHI Newcastle 1.
70 OHI Newcastle 9.
71 OHI Newcastle 1.
72 OHI Newcastle 1.
73 OHI Newcastle 1.
74 OHI Newcastle 2.
75 OHI Newcastle 10.
76 OHI Newcastle 11.
77 OHI Newcastle 12.
78 OHI Newcastle 12.
79 OHI Newcastle 8.
80 Newcastle upon Tyne Town Moor Act 1988, recitals.
81 Ibid. s.4.
82 Ibid. s.6(2).
83 Ibid. s.6(7).
84 Ibid. ss. 5 (3)–(5).
85 Ibid. s.8(1).
86 Ibid. s.10(2).
87 Ibid. s.10(3).
88 Ibid. s.10(4).
89 Ibid. s.9 (2).
90 Ibid. s. 9 (3). Leases of more than 21 but less than 99 years can be granted with the permission of the Charity Commission. The terms for advertising auctions and the terms of prospective leases to be auctioned are set out in s. 9(6) ibid.
91 See ibid. s. 7.
92 See Bye-laws for the Regulation of Cycling and Vehicles on the Town Moor (1999): see https://www.newcastle.gov.uk/local-government/byelaws-regulation-cycling-and-vehicles-town-moor (accessed 15 October 2021).
93 See ibid. s 8(5).
94 Recreational facilities or allotments can be permitted for periods up to seven years, although this can be renewed: ibid. s. 8 (ii).
95 https://www.newcastle.gov.uk/sites/default/files/2019-06/152%20Sites%20of%20Local%20Conservation%20Interest%20DAP%20Evidence.pdf
96 Ibid. at p. 57.
97 Ibid. at p. 56.
98 Semi-structured interview, 21 May 2021.
99 https://www.newcastle.gov.uk/sites/default/files/2019-06/152%20Sites%20of%20Local%20Conservation%20Interest%20DAP%20Evidence.pdf at p. 56.

100 Semi-structured interview, 21 May 2021.
101 Historic Newcastle allotments 'wiped clean of life', BBC News, 5 February 2018. https://www.bbc.co.uk/news/uk-england-tyne-42413754 (Accessed: 13 December 2021).
102 OHI Newcastle 12.
103 OHI Newcastle 12.
104 OHI Newcastle 10.
105 OHI Newcastle 10.
106 OHI Newcastle 8.
107 OHI Newcastle 1.
108 OHI Newcastle 5.
109 OHI Newcastle 6.
110 OHI Newcastle 5.
111 OHI Newcastle 5.
112 OHI Newcastle 10.
113 OHI Newcastle 8.
114 OHI Newcastle 2.
115 For further details on the participatory research and learning activities carried out for this project, see Chapter 9.
116 Newcastle City Council *Sites of Local Conservation Interest (SLCI) Reports, Development Allocations Plan Evidence* (2013), p. 56. Available at: https://www.newcastle.gov.uk/sites/default/files/201906/152SitesofLocalConservationInterestDAPEvidence.pdf
117 Participant Comment, Newcastle University APL, Student Project Intervention. *Sheltered Seating* by Emma Itu. 2021.
118 Ibid.
119 Ibid.
120 Ibid.
121 Ibid.
122 Ibid.
123 Ibid.
124 Ibid.
125 Newcastle University APL, Student Project Participatory Pack 5. 2021.
126 Participant Comment, Newcastle University APL, Student Project Intervention. *Sheltered Seating* by Emma Itu. 2021.
127 Ibid.
128 Newcastle University APL, Student Project Participatory Pack 5. 2021.
129 Newcastle University APL, Student Design Project. Remembering the Witches by Darcy Morse. 2021.
130 Newcastle University APL, Student Project Participatory Pack 1. 2021.
131 Participant Comment. Newcastle University APL, Student Participatory Intervention. Casting More Light by Quitterie D'Harcourt. 2021.
132 Ibid.
133 Ibid.
134 Quitterie D'Harcourt. Urban Commons – Group 1 Portfolio (Newcastle University APL, 2021), p. 94.
135 See UN Environment, Convention on Migratory Species. Thirteenth Meeting of the Conference of the Parties, Gandhinagar, India. 17–22 February 2020. UNEP/CMS/COP13/ Doc.26.4.9.1/Rev.1. Available at: https://www.cms.int/en/document/light-pollution-guidelines-wildlife-including-marine-turtles-seabirds-and-migratory-0
136 All-Party Parliamentary Group for UN Women. (2021). *Prevalence and reporting of sexual harassment in UK public spaces.* YouGov. p. 6. Available at: https://www.unwomenuk.org/site/wp-content/uploads/2021/03/APPG-UN-Women-Sexual-Harassment-Report_Updated.pdf
137 Ibid., p. 7.

138 Newcastle University APL, Student Participatory Intervention. *The Future* by Sarah Bird. 2021.
139 Participant Comment, Newcastle University APL, Student Participatory Intervention. *Seeds of Connection* by Rachel. 2021.
140 Participant Comment, Newcastle University APL, Student Participatory Intervention. *The Future* by Sarah Bird. 2021.
141 Ibid.
142 Newcastle University APL, 3rd Year Student Project Intervention. *Bird Boxes* by Diana Mihailova. 2021.
143 Participant Comment, Newcastle University APL, Student Participatory Intervention. *More Trees?* by Thomas Paramor. 2021.
144 Participant Comment, Newcastle University APL, Student Participatory Intervention. *Seeds of Connection.* 2021.
145 Participant Comment, Newcastle University APL, Student Project Participatory Pack 1. 2021.
146 Newcastle University APL, Student Participatory Intervention. *Cleansing Newcastle's Past* by Jeremy Bidwell. 2021.
147 Ibid.
148 Association for Public Service Excellence (2022). *State of the Market Survey 2022. Local Authority Allotment Services.* Available at: https://apse.org.uk/index.cfm/apse/members-area/briefings/2022/22-33-state-of-the-market-allotments/
149 Newcastle University APL, Student Participatory Intervention. Cleansing Newcastle's Past by Jeremy Bidwell. 2021.
150 Ibid.

3 Mousehold Heath, Norwich

Introduction

Location

Mousehold Heath is a 184-acre heathland, woodland and recreational open space within the administrative boundary of Norwich in eastern England. Although the Heath is currently situated in a predominantly urban zone, historically the site was much larger (c.6000 acres), stretching from the medieval city of Norwich to the Norfolk Broads. Consequently, while in some respects it might look today like an archetypal urban common, it, in fact, presents a complex case study which has been subject to change over time and has long been a site of conflict and contested memories.

Topography/Features

Various natural and man-made features survive as remnants of past human use across Mousehold Heath. By the eleventh century, much of the original ancient woodland on the Heath had reduced and regrowth was prevented by animal grazing. Although little remains of this earlier landscape use, the Long Valley, a naturally occurring dry channel, is a useful reminder. The valley is one of many former sheep walks across the Heath that provided communication and trade connections between Norwich and the surrounding villages. In contrast, the remains of sand, gravel and clay quarries, which increased in production from the sixteenth century, have made a longer-lasting impression. Most of these quarries are now subsumed within woodland regrowth, giving the impression of an undulating landscape, but they were an essential part of the local economy with the Stone Pit Company quarry on Valley Drive extracting until the mid-twentieth century. Some of these topographical features have found new uses amongst current users of the space such as dog walkers and BMX bikers.

Narrative

The early history of Mousehold Heath is difficult to establish, but the earliest definitive reference to the name Mousehold or "Musholt" dates to 1315.[1] As cartographic

DOI: 10.4324/9781003204558-4

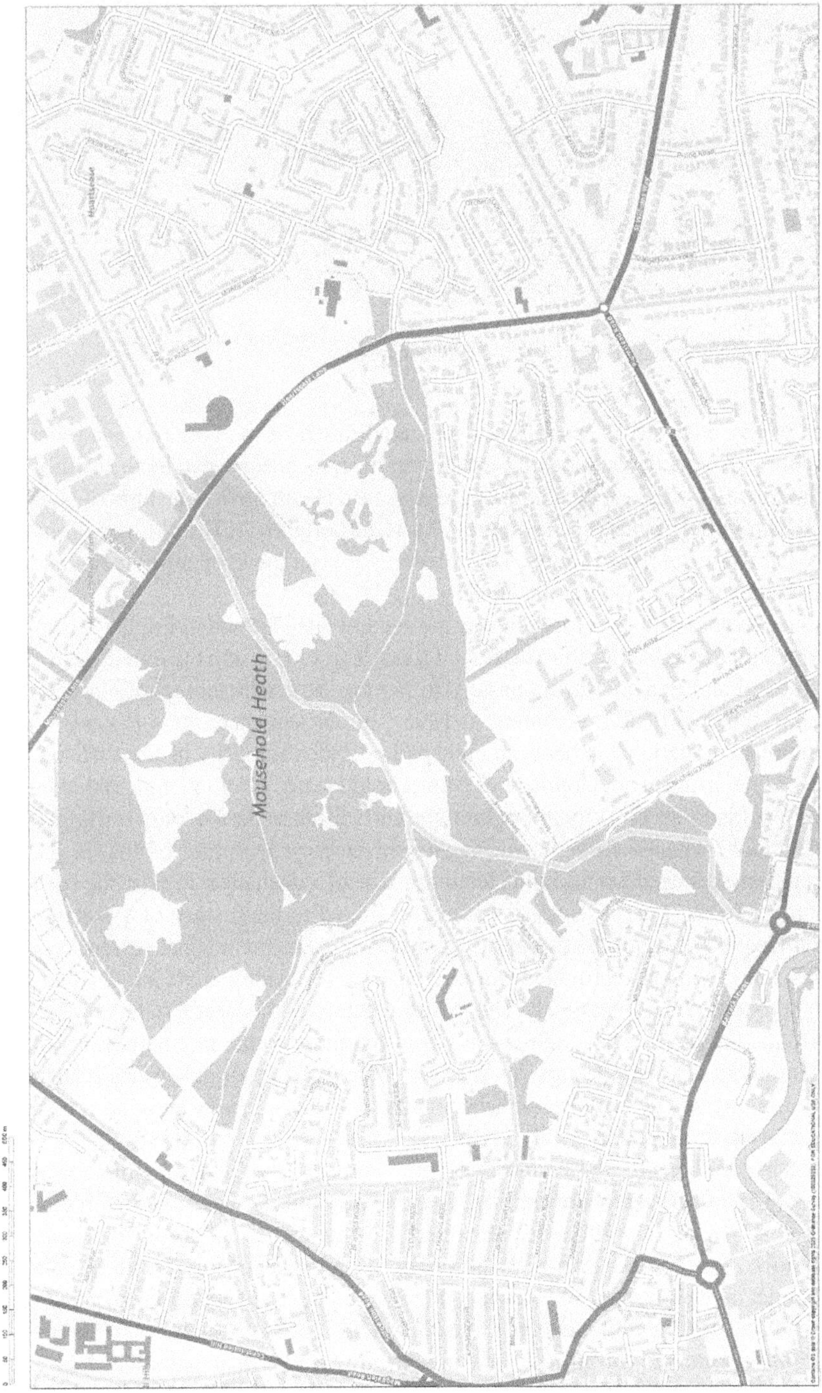

Figure 3.1 Map of Mousehold Heath.

Contains public sector information licensed under the Open Government Licence v3.0.

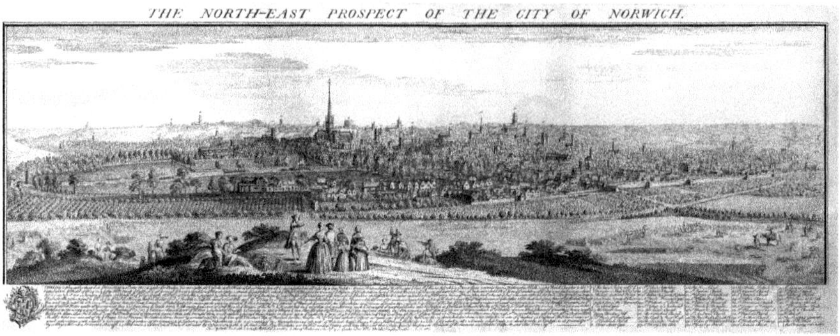

Figure 3.2 The North-East Prospect of the City of Norwich.

and settlement analysis of Norfolk has indicated much of the common land that existed in Norfolk in the eighteenth century was already in existence by the Middle Ages; in this regard, the establishment of Mousehold Heath on the outskirts of Norwich fits similar trends across the county.[2] More importantly, it was this proximity to Norwich that established a close relationship between the economic, social and political history of the city and the Heath.

By the eleventh century, Norwich was the second largest town in England. Norman expansion included the construction of a castle in c.1067 and the founding of a new cathedral and Benedictine monastery following the establishment of Norwich as a centre for episcopal government in 1095.[3] In consequence, there has been a well-established relationship between civic and religious authority in Norwich that connects to the historic development of Mousehold Heath. Norwich maintained its status as England's second city throughout the early modern period, only starting to drop in national statistics from the mid-eighteenth century. For much of this period, local communities used Mousehold Heath as a site of sustenance. In fact, the built environment of this part of Norfolk is directly linked to the history of extraction from the Heath by way of the use of raw materials in settlement construction.

Mousehold Heath also provided extramural space to accommodate large gatherings and protests close to Norwich. The elevated nature of the Heath, relative to the city, provided a strategic position and a visual reminder to the city authorities of the potential threats these groups could pose. Of course, not all gatherings related to protest; alongside agricultural improvements, recreation increasingly became a driving motivator for change.

In the late eighteenth and early nineteenth centuries, the landscape history of Mousehold Heath changed irreversibly when it was split up under various Inclosure Acts. By 1838, only 190 acres remained. Despite continued use of the Heath for livestock grazing and extraction purposes by the residents of Pockthorpe, a small hamlet on the outskirts of Norwich, there were increased pressures to convert and protect the site. In consequence, Mousehold Heath became the focus of deeply contentious arguments during the late nineteenth century that culminated in a lengthy legal battle that determined the rights of Norwich Corporation to establish the Heath as a "people's park". Despite development pressures, Mousehold Heath has continued to act in this capacity to the present day, providing recreational space for a diverse set of user groups.[4]

Historic Mousehold Heath: Governance and Property Rights

Henry I granted the manor of Thorpe to the bishopric of Norwich in 1101 to meet the expense of building Norwich Cathedral. Between 1101 and 1290, the manor was divided into Greater (or Bishops) Thorpe and Little Thorpe (or Pockthorpe). Edward VI granted a charter to the Dean and Chapter of Norwich in 1547 that stipulated their rights over the manor of Pockthorpe, which included a grant of "all the King's moors, marshes, void grounds, sheep walks, and foldages there".[5] These arrangements, as to the manorial ownership, land use and property rights on the Heath, remained largely intact until the second half of the nineteenth century.

The legal status of Mousehold Heath, and the property rights that Norwich residents could claim over it, was a highly charged issue in the Victorian period. The emergence of the Heath from the medieval manorial system and its transformation into the vibrant recreational green space that we see today saw lengthy and complex legal proceedings before the Victorian inclosure commissioners and then the courts. This culminated in Chancery litigation in 1883, leading to a decision by Mr Justice Chitty which confirmed the title of Norwich Corporation to the Heath. Some of the issues raised in this litigation – for example, the ability (or otherwise) of community action to generate property rights, and the relevance of long usage and occupation for the informal creation of rights in land – are considered further in Chapter 8.

Property Rights and Land Use

The Heath was formerly waste of the manors of Thorpe, the title to which was until 1869 vested in the Dean and Chapter of Norwich Cathedral. The process by which the Heath was transformed from manorial waste – common land with grazing rights for sheep and cattle depastured by Norwich residents – into an entirely different form of community asset owned by the Norwich Corporation was both lengthy and hotly contested. The principal dispute concerned the rights (or otherwise) of Pockthorpe residents to extract gravel, sand and clay from pits dug on the Heath for the purpose of brickmaking. The origins of this practice are obscure, but it was a well-established activity that had been carried on since at least the sixteenth century.[6]

In the sixteenth and seventeenth centuries, gravel and clay extraction seems to have been supervised closely by the Cathedral Chapter through the Pockthorpe manor court.[7] Manorial court supervision of the practice seems to have lapsed during the eighteenth century, and by the 1840s, the practice of digging sand and gravel was instead being collectively organised by the local community. The Pockthorpe committee, established in 1844, was an independent example of community action, unsanctioned and separate from the "official" channels of community organisation represented by the Norwich Corporation and the Dean and Chapter of Norwich Cathedral (who in 1844 were still lords of the manor of Pockthorpe).

The transfer of ownership to Norwich Corporation and the transformation of Mousehold Heath into a managed "people's park" was by no means a simple process: ownership of the Heath was contested by the Pockthorpe community and proposals to reinvent it as a space for community recreation met with strong resistance.[8]

The sequence of events leading to the Chancery litigation of 1883 started in 1866, when the Dean and Chapter of Norwich Cathedral agreed to a scheme with the ecclesiastical commissioners for the creation of pleasure grounds with recreational public access at Mousehold. This was followed on 17 March 1869 by an Order in Council which vested the freehold title to Mousehold Heath in the ecclesiastical commissioners. This ended the legal involvement of the cathedral chapter in the subsequent development of the Heath, extinguishing their title as lords of the manor over the waste of the manors of Thorpe. The Public Health Act of 1875 subsequently empowered local authorities to create and hold land as "pleasure" or recreation grounds. This was an important step in the realisation of the Victorian ideal for the provision of public parks as orderly green spaces for open air recreation and exercise, and the powers were used across England and Wales to create many of the well-ordered urban parks that we still see today. Norwich was no exception. In July 1880, 184 acres (and 35 perches) of Mousehold Heath were conveyed to the Norwich Corporation to hold as a recreation ground "subject [only] to rights of common, easements and other rights".[9]

Meanwhile, the activities of the Pockthorpe community on Mousehold Heath had continued. Gravel and clay were extensively removed and used for brickmaking.[10] After it acquired the legal title, the Norwich Corporation reached an accommodation with some of the brickmakers, giving them tenancies to temporarily continue their operations. Nevertheless, those entering into tenancies were threatened and their access to and from the brickfields impeded; the roads onto the Heath were also blocked with gates to prevent brickmakers from neighbouring parishes such as Sprowston from accessing the clay and gravel pits to source their own supplies. This led to prosecutions under the Highways Acts before the Norwich magistrates on several occasions in the early 1880s.[11]

Norwich Corporation presented a draft scheme for Mousehold Heath to the inclosure commissioners in 1881. This was challenged by the Pockthorpe committee, who entered a memorandum claiming that they (not the Corporation) held title to the land based on their use of it for sand and gravel extraction "from time immemorial". This halted the scheme's approval, as the commissioners had no power to make rulings on matters of title to land. The Corporation issued legal proceedings in 1881 before the Court of Chancery seeking an injunction, both to prevent the Pockthorpe committee interfering with the commissioners' hearings and to prevent them interfering with the Corporation's tenants who were using the Heath. In July 1881, the Court of Chancery issued an injunction preventing the removal of gravel or soil from the Heath and prohibiting Pockthorpe residents from blocking roads leading onto the Heath.[12] This was followed by another action in Chancery in June 1883, in which Justice Chitty upheld the Corporation's legal title to the Heath.

The legal argument put forward by the Pockthorpe brickmakers was ingenious, but ultimately failed to satisfactorily address several fundamental principles of English property law. Unfortunately for the Pockthorpe brickmakers, the right to take gravel, clay and soil from someone else's land has never been recognised as a legitimate right of common.[13] They were therefore unable to claim that their "rights" were preserved either by the terms of the 1880 conveyance to the Corporation or by the Public Health Act 1875.[14] Most common rights to take the soil's produce are characterised as *estovers*: these typically attach to houses or cottages adjoining a common and give the holder the right to take necessary produce for food, fuel, timber or animal bedding. They are not rights that can be used for commercial purposes. The rights claimed by the Pockthorpe residents here were essentially needed to support small-scale industrial activities – to take clay and gravel for the manufacture of bricks. This was a type of activity for which long usage could not create property rights by prescription under the common law.

The Pockthorpe committee therefore adopted an alternative strategy and claimed that they had acquired ownership of the land itself by adverse possession; or rather that the title to the land was vested in the trustees of a charitable trust for the relief of poverty in Pockthorpe, over which they had control. The origins of the "trust" were traced back to an open vestry meeting of the Pockthorpe parish on 23 May 1844, which had elected a secretary and treasurer to oversee the collection of payments and their distribution as vouchers to residents. The de facto control of access to the brickfields by the "trustees" under this arrangement had largely gone unchallenged by the Cathedral authorities, who as lords of the manor had for many years held the "paper" legal title to the Heath. This led to some criticism of the church authorities in the court proceedings, not least from Justice Chitty himself who commented that "what has happened is to some extent due to the laxity, if not the supineness, of the administration in later times of the Dean and Chapter of Norwich".[15]

The key problem for the Pockthorpe community's claim was that to constitute a trust the legal title to the land must vest in an individual or corporation. Justice Chitty observed that:

> to gain an adverse title under the Statute of Limitations the possession must not be in one man one day, and in another [on] another. There was no legal possession in these persons at all. It was in a changing and fluctuating body of persons who were not acting on their own account.[16]

Legal title had not been vested in specific named members of the community who could act as the trustees, and it could not vest in "officers" elected each year to oversee the arrangements for the collection and disbursement of fees, as these would change over time. On the other hand, however negligent they may have been in exerting control over the Heath, the paper title of the Cathedral Chapter and (then) Norwich Corporation was proved to the court's satisfaction. The Corporation's lawyers produced deeds and charters going back ultimately to the Norman Conquest to show an unbroken line of ownership of the Heath.

Governance of the Heath

Following their victory in Chancery, the Corporation acted swiftly to finalise their plans for the transformation of the Heath. In December 1883, the Land Commissioners for England[17] certified a scheme for the development of Mousehold Heath as pleasure grounds under the City of Norwich Act 1867. This was implemented by the City of Norwich Mousehold Heath Scheme Confirmation Act 1884, which set out a management scheme for the Heath and provided for the appointment of the first body of Conservators, of whom there were to be 12: nine nominated by the Corporation, two by the inhabitants of Pockthorpe and one by the Commissioner of Public Works. The total sum spent on the maintenance and upkeep of the Heath in any one year was fixed at a maximum rate of 1/2 d in the £1 and no expense greater than £25 was to be made without the consent of the Norwich Corporation.

The terms of the 1884 Act reflect, and were shaped by, the history and land use of the Heath in the Victorian period. Interestingly, it was still regarded as a "common" in the accepted legal sense: the Act refers to it as "divers waste lands and a common called Mousehold heath…"[18] A certificate issued by the chief clerk to the Corporation in July 1882 certified that the only rights of common lawfully subsisting at this point were rights of pasturage for sheep and cattle *levant and couchant*. These rights were not subsequently registered under the Commons Registration Act 1965, and the Heath ceased to be a "common" in the narrow legal sense once the 1965 Act registrations became final in 1970. From this point on, it is probably better referred to as "community" property: land owned by the Corporation but subject to open recreational access rights for the people of Norwich.[19]

The Pockthorpe dispute shaped the terms of the scheme and the bye-laws subsequently introduced to govern access to and use of the Heath and its resources. These were redolent of the Victorian ideal of "managed" recreational use by the public, with a focus on the prevention of crime and of improper and socially disruptive behaviour – facets which place the legislation firmly within the scope of the Victorian parks movement. The corporation was to "take all lawful means to prevent the *continuance* of trespass, nuisance and unlawful acts"[20] on the Heath and to maintain it free from any encroachments; it was moreover deemed to be a place of "public resort" where the powers of the police regarding public order were to apply. The Corporation's role was defined in the scheme as one ensuring the "proper" management of the Heath and the preservation of "orderly" conduct.

The Pockthorpe dispute itself may be long forgotten, but the legacy of its origins in the industrial history of Norwich lives on in the landscape of the Heath. As noted above, its hollows and pits are a prominent feature of the Heath and are in large part what makes it such an interesting and valued landscape for the local community. Contemporary users have commented in oral history interviews on the "interesting" and "unique" nature of the landscape, which makes it both special and central to the identity of Norwich as a modern city. The contemporary notions of "belonging" and "identity" associated with the Heath are explored further below (see Chapter 8).

The Changing Space of the Common

The history of Mousehold Heath is a familiar story of loss. Despite centuries of human use, ultimately the physical space of the Heath was subject to episodes of gradual and dramatic erosion. There is a difference, therefore, between the landscape area now designated as "legal" Mousehold and the former, much larger expanse of Mousehold Heath. The mutability of this boundary was also evident in the response of local users to activities instigated as part of the Wastes and Strays project, which revealed that individual and community perceptions of where Mousehold Heath begins and ends could be very fluid indeed. Interestingly, and in contrast with other English urban commons, it was responses to changing agricultural methods that started the process of encroachment, followed by increased pressure on the provision of recreational resources and urbanisation.

William Faden's Map of Norfolk (1797) (http:fadensmapofnorfolk.co.uk) is accepted as offering a good indication of the extent of Norfolk's common land before parliamentary enclosure.[21] As Tom Williamson and Andrew McNair have observed, the irony of Faden's map was that within the next three decades, the vast majority of common land that he depicted would be lost to agricultural "improvement".[22] In some instances, enclosure of Norfolk's commons had already taken place, but the fact that Faden's depiction of Mousehold Heath is consistent with earlier – sixteenth century – maps means that we can be reasonably confident that the Heath was about 6,000 acres. If we consider that Faden calculated Norfolk's heaths to be c.16,000 acres, Mousehold Heath represented a significant share of the county's heathland landscape. Furthermore, with a circumference of 22 miles, Mousehold Heath directly contributed to the local economies of several Norfolk parishes, including Blofield, Great and Little Plumstead, Norwich, Postwick with Witton, Rackheath, Salhouse, Sprowston and Thorpe St Andrew.

The threat of enclosure was an issue that impacted several parts of Norfolk and Suffolk during the early modern period. Most notable, for Mousehold Heath, was the site's strategic value for Robert Kett and his followers during the 1549 protest against countrywide piecemeal enclosure. However, Mousehold Heath was not directly threatened with enclosure during these protests. Instead, prior to the early nineteenth century, only small losses had taken place through local clearance and encroachments that were usually only one or two acres surrounding "New Farms".[23] It was really during the eighteenth century that discussion concerning the benefits of enclosure increased. Authors, and cartographers such as Faden, increasingly emphasised the benefits of enclosure to appeal to their landowning readership. John Wagstaffe's proposal to enclose Mousehold in 1792 provides a useful example.[24] In some respects, Wagstaffe was progressive in his suggestion that the poor and their descendants should expect compensation for the loss. But he also placed emphasis on the greater benefit that would be achieved from cultivation – with wildly inaccurate estimations that the Heath comprised 20,000 acres, Wagstaffe estimated that 10,000 people would be uplifted by the Heath's enclosure.[25] In addition, Wagstaffe suggested that the inhabitants of the city of Norwich would benefit if at least a third of Pockthorpe Heath – that part of Mousehold within the Norwich boundary – was converted into pleasure grounds for exercise.[26]

The ideas that emerged during the eighteenth century resulted in transformational landform change to Mousehold Heath in the early nineteenth century. From 1801, parliamentary enclosure was more significant than anything that people had experienced in Norfolk previously and the enclosure of the greater extent of Mousehold Heath – outside of the Norwich boundary – would have produced a dramatically different landscape for those communities that bordered the Heath and used the sheep walks to connect with one another.[27] Within a generation, romanticised accounts such as the anonymous "Tales of Mousehold" (1835) highlighted the personal impacts of landscape change:[28]

> During the interval, they had been enclosed, and the barrier of long fences, and the vigilance of rigid and tenacious proprietors, excluded him from the greater part of those favourite haunts, which had so often witnessed his eager pursuit of juvenile sports, or the meditative rambles of more serious hours.[29]

The personal and physical impact of the change cannot be overstated. The 1801 Act alone is thought to have enclosed approximately 1,000 acres within a single phase, and by 1838, only half of the Heath at the Norwich end remained (c.190 acres).[30] It was this area that was transformed into recreational space via the complex and contested process and legal proceedings described above.

For much of the eighteenth century and up until 1866, the Dean and Chapter of Norwich leased out parcels of the Heath as part of their land management strategy. In effect, as owners, the Dean and Chapter's Office was aware that the physical geography of Mousehold Heath was changing as a result of mineral extraction. Furthermore, despite extensive enclosure of the larger portion of Mousehold Heath, the Dean and Chapter's Office had made no attempt to erect or maintain fences with bordering land.[31] Mr Justice Chitty summarised the root of the issue during his 1881 judgement in favour of the Norwich Corporation. Mousehold Heath had become "that kind of land upon which persons stray – and particularly those who inhabit the neighbourhood do something more than stray – they obtain by degrees some notion that it is their right and their property".[32] This can be seen as a prefiguring of the perception of contemporary users that through acts of commoning, they can take back or informally "unenclose" parts of the Heath.

In some respects, it may have been the enclosure of the majority portion of Mousehold Heath that focused issues concerning use and oversight to the remaining Norwich portion. The Dean and Chapter's Office increasingly pursued the idea of relinquishing parts of the Heath to Norwich Corporation, despite, or perhaps because of, the attempted management by the Pockthorpe committee from 1844. In 1849, the Dean and Chapter offered a portion of the Heath (without purchase) for use as a cemetery – this offer was renewed in 1853 and again in 1855.[33] Likewise, the Corporation also made enquiries in 1855 regarding the cost of purchasing an area of the Heath to build a school in Pockthorpe for the poor children in the hamlet.[34] It is clear that after 1838, the Dean and Chapter's administration of its property was threatened in ways that it had not been previously. This may be due to the pressures associated with exerting property rights over a much smaller landmass and, as a consequence, over far fewer shared resources.

The Chapter canons were well aware of the problems that they faced. In 1863, when they attempted to cut turf on Mousehold Heath, they were forcibly prevented from doing so by the local community.[35] George Pellaw, writing on behalf of the canons in November 1864, was blunt – as administrators, the Dean and Chapter's Office had not exerted their rights for several years and in consequence they could only expect to regain them through legal measures, which would be costly and unpopular.[36] During the ongoing discussions, only one voice counselled for the continued patronage of the Pockthorpe residents, which could have been achieved by setting aside a small part of the Heath for enclosure and lease. The greater majority resolved to make use of the upcoming general election – when politicians were "disposed to become public benefactors to the inhabitants of Cities and Boroughs" – and in a shrewd manoeuvre offer all the interests and rights in the unenclosed portion of the Heath to the Corporation for the purpose of adapting it into a "people's park".[37] They did so knowing that they were "offering the City a lawsuit", but could cite the success of similar ventures such as Wimbledon Park, which had been developed from common land during the 1840s.[38] The Corporation accepted the offer during a meeting of the Council of Norwich in April 1866 and took possession of a lease in December of the same year.[39] Already by 1867, a draft scheme for the conversion of the Heath was finalised, but, as described above, a lengthy legal battle followed between the Corporation and representatives of the Pockthorpe residents.[40]

In terms of the Heath's physical geography, it is important to acknowledge that although the Mousehold Heath Scheme of 1884 facilitated the protection of "legal" Mousehold by a group of Conservators, this was, and still is, an adapting landscape. From the outset, Mousehold Heath continued to be used as an economic resource for the local population – albeit in a severely reduced capacity – which changed or broke up the previously unenclosed space. For example, although significantly reduced, mineral extraction continued by the Mousehold Heath Stone Pit Company into the twentieth century, and the impact of these works is still evident in the landscape of the Heath today. In addition, the Heath was broken up by the insertion of new roads, such as Gurney Road, which provided better access across the space and much needed local employment opportunities as a form of relief for the poor.[41]

Of greater significance has been the subsequent incorporation of the Heath into the built environment of Norwich, as urbanisation became an inevitable process with population expansion. There were still sections of Heath immediately adjacent to "legal" Mousehold that became cut off and subsequently urbanised from the late nineteenth century. Figure 3.3 reveals how Mousehold Heath was incorporated into the city scheme.

Four phases of abutment can be observed between the 1880s and the present in Figure 3.3. Between 1880 and 1914, parts of the adjacent Heath marked in black with grey spots were privately developed for housing, and two sections on the southern boundary were obtained by the War Office to build Britannia Barracks

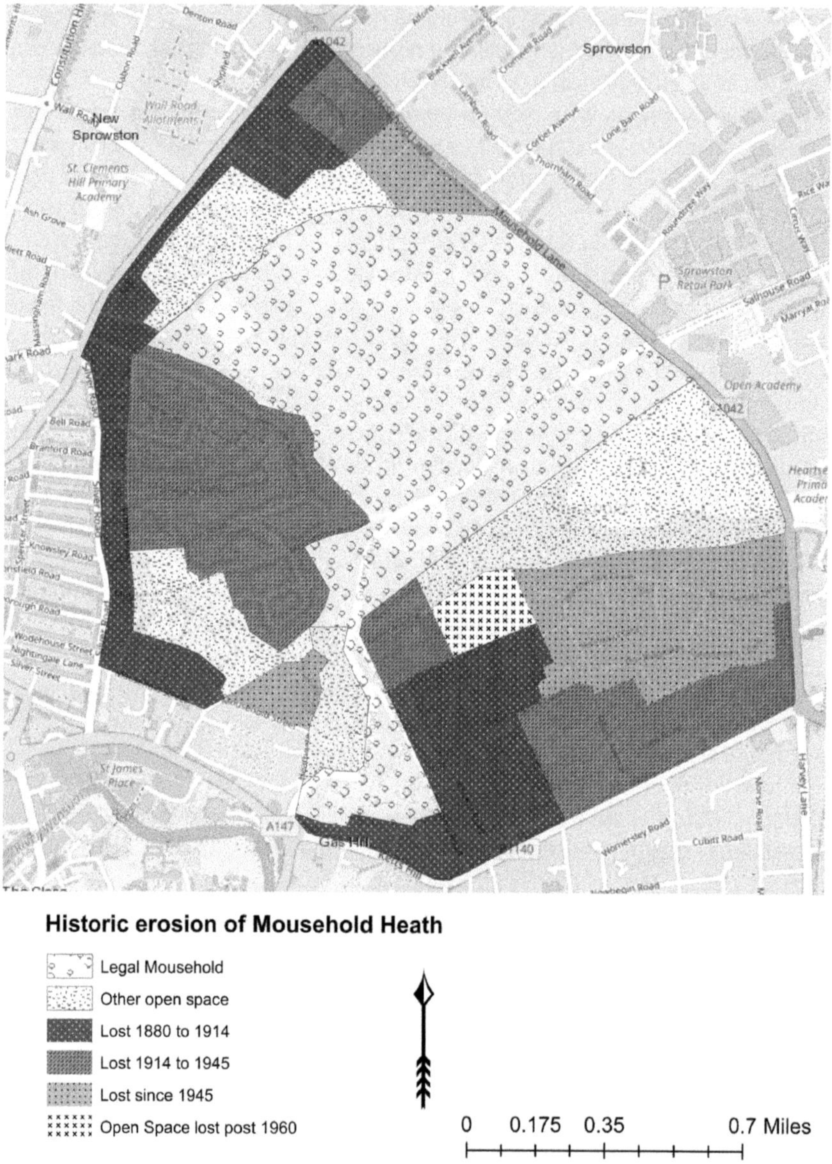

Figure 3.3 Map depicting the erosion of Mousehold Heath over time.

(1884) and by the Home Office to build a prison (1885). Both buildings are now so significantly incorporated into the Mousehold landscape that they form the basis of important oral history connections in their own right. In addition, there were two major phases of private housing development (dark grey with black lines and mid grey with black spots) – the first during the interwar period enclosed parts of the former adjacent heathland to the south and west, and the second, post-war phase made further encroachments into the green space between the prison and Mousehold Lane. Finally, a small extension of Norwich prison has resulted in further loss to the adjacent green space post-1960, a loss that was advised against in order to unify Mousehold in 1964.[42] The consequence of these changes are the continual adaptation in character of the landscape that surrounds "legal" Mousehold, and, more importantly, fears for the survival of the surrounding green space that has increasingly limited capacity to act as a protective zone.

The Urban Common in Social and Political Context

Just as what survives of Mousehold Heath is only a remnant of the original common, so there is a discrepancy in the use of the space between the overt and the hidden, between the remembered and the forgotten. The recreational use that dominates today has a long history, but other uses have also left their mark on the landscape and in the documentary record.

Recreation

The Heath has been a venue for formal and informal sporting contests since at least the eighteenth century. In October 1767, waterman Robert George fought John Todd, a worsted weaver, in a boxing match on the Heath. Such contests continued into the nineteenth century.[43] Racing was another popular sport in the early nineteenth century. In November 1824, an arrangement was made with a local landowner (Colonel Harvey) to turn a portion of his estate on Mousehold into a race course in the week before the assizes.[44] The use of the Heath for this purpose was, however, sporadic. In 1838, it was announced that the races would be reintroduced to celebrate the coronation of Queen Victoria:

> Races at Norwich may be considered almost a novelty, for although there has been a fine open space of ground on Mousehold Heath bearing the name of the Course, no use, or at least so little use has been made of it for many years, that a large proportion of the inhabitants do not even know where it is.[45]

A subsequent article drew attention to the popularity of the event, suggesting that it had attracted 30,000–50,000 people.[46]

Private pursuits also took place on the Heath, as evidenced by Rosa Howes's letter inviting her friend Dollie Freeman to have tea with her at "the garden Mousehold".[47] Artists were particularly drawn to the Heath. In November 1841, John Sell Cotman, the English marine and landscape painter, wrote to the art collector Dawson Turner:

> I galloped over Mousehold Heath … but was obliged to stop and sketch a magnificent scene on the top of the Hill leading down to Col. Harvey of Trees and gravel Pit. But Norfolk is full of such scenes, oh rare and beautiful Norfolk.[48]

Cotman was a leading figure in the Norwich school of artists that flourished in the early nineteenth century.[49] These artists – who specialised in landscape paintings, especially tranquil rural scenes – painted various locations around Norwich, including Mousehold Heath. Their depictions contributed to the idealisation and mythologising of the landscape.

Different recreational activities did not always sit comfortably alongside each other. A correspondent to the *Norfolk Chronicle* in 1855 complained that: "the promenade of the adult is frequently disagreeably checked by finding himself in rather too close propinquity to the noxious viper, and the gambols of the young".[50]

The 'People's Park' and the Erasure of – and Conflict Over– Earlier Common Rights

It was in part to protect the Heath as a recreational space, and to control the sorts of activities that took place there, that proposals were made to convert it into a "people's park". The plan drawn up in 1865 stated that the Conservators would oversee improvements to the Heath, paying attention to planting, roads and seating and sports grounds. In prioritising recreation, this initiative served to erase other uses of the space. New bye-laws not only restricted the kind of games that could be played – with explicit strictures against fighting, bad language and disorderly behaviour – but they also rescinded earlier common rights. The bye-laws served to prevent inhabitants from cutting or digging turf; from cutting trees and collecting wood, gorse and heather; and from exercising horses. In acknowledgement of the recent dispute, they also explicitly forbade the digging and selling of sand and gravel, as well as legitimising "the exclusion and removal of gamblers, card sharpers, gipsies [sic], squatters, vagrants", and empowering the Corporation to remove "any idle or disorderly persons" from the Heath.[51]

While there are no grazing rights on Mousehold today, there are plenty of signs of this activity in the documentary record and the landscape. From the sixteenth century, inhabitants of 11 parishes had the right to graze animals on the common. Neighbouring lords also claimed foldcourse (sheep grazing) rights over portions of the Heath. These grazing rights were accompanied by other common rights, including the right belonging to the "poor folks of Norwich and of Pockthorp" to gather furze and flags for fuel.[52] There is evidence of mineral extraction dating back at least to the sixteenth century. The local Historic Environment Record notes the existence of sand, gravel and clay quarries of various sizes across the Heath.[53] Also evident are the remains of the kilns, which were flattened when the Heath was converted into a park.[54] As well as leaving evidence in the landscape, extraction sites also appear on historic maps and conflicts over them are recorded in the archives. The digging of minerals was, as demonstrated above, a source of conflict between local residents and the Dean and Chapter, which took steps to protect its rights "in the Soil and Herbage of Pockthorpe heath and to prevent the indiscriminate digging and taking away of the same". In 1801, the overseer James Houghton was tasked with ensuring that nobody took gravel or stone away without the acknowledgement of the Dean and Chapter, with further measures introduced in 1828.[55] When the Pockthorpe committee was established at an open meeting of the local vestry in 1844, it resolved to make a charge for the right to remove gravel and clay from the common. To this end, the committee was tasked with: deciding which areas were to be opened as gravel, stone or clay pits; enforcing a standard list of charges for taking materials; ensuring that dues were properly collected; and distributing the profits among the local poor each Christmas in the form of vouchers (tickets) redeemable at local stores. In an attempt to prevent the exploitation of the Heath by outsiders, the committee introduced lower rates for materials taken by inhabitants and required local labourers to be employed to dig and cart, provoking conflict with brickmakers from neighbouring areas. Ultimately, as described above, this sophisticated example of local community organisation failed to secure legally recognised rights to property or to the resources of the Heath in the later Chancery litigation of 1883.

Military Usage, Crime, Illness and Death

Another activity that is important to the history of the Heath, but which is not always obvious to the untrained observer, is its deployment for military purposes. The area known as The Heights retains a six-metre high earthwork that formed a set of butts for shooting practice that was in operation prior to 1863.[56] During World War II, Mousehold was used for military training, incorporating an area where trench digging could be practised, weapons pits and a rifle range. It was also a location for defensive structures, including barrage balloons, anti-aircraft guns and a dummy airfield[57] as well as being the site of a prisoner of war camp. No doubt because of its defensive role, the area suffered considerable bomb damage. In one place, at least six bomb craters have been detected through the National Mapping Programme analysis of aerial photographs.[58] The Heath was also the site of two plane crashes.

The location of Mousehold, on the edge of the city, meant that it could be used to hide illness, crime and even death. Lazar House was a leper hospital from the twelfth to the sixteenth centuries.[59] It was subsequently converted into almshouses and then into a library. The mixed flint and brick construction is a good example of the use of materials from the Heath in the local built environment. Among various examples of Mousehold being used for criminal activities, a report from 1835 is particularly telling. The police were chasing three suspected thieves, but when the suspects ran onto the Heath, the police ceased their pursuit, claiming that the area was beyond their jurisdiction.[60] Contraband was also hidden on the Heath. In one case, in 1825, a large quantity of cheese was found in a gravel pit.[61] Those pits could also cause accidents, as a number of coroners' inquests from the eighteenth and nineteenth centuries testify. Christopher Custance's fall into a pit was said to be so violent that he received immediate bruising and spat up "great quantities of blood".[62] David Spencer died in similar circumstances in 1751 when he fell from the edge of the gravel pit where he was working.[63] Those working in the pits could also be suffocated when sections above gave way. Robert Burgess (1760), James

Nudd and Robert Barker (1823) all died in this way.[64] Murders occasionally took place on the Heath. In 1758, James Rippen sustained a fatal bullet wound to the arm while watching military drills. Since the hand that fired the shot was unknown, the coroner recorded a verdict of "wilful murder".[65] In 1822, workers discovered human remains in a sand pit. A witness claimed they belonged to a young woman who had been robbed and murdered two years earlier. However, the surgeon who dealt with the remains insisted that the bones were those of a man, leading to the recording of a verdict of "wilful murder against some person or persons unknown".[66] The Heath was also used at various times for burials. Medieval human remains were found on Gilman Road and others from the eighteenth or nineteenth centuries were discovered on Roundtree Way.[67] The latter may have been from the Romany community, who were known to enact burials on the Heath.[68]

Remains, Memories and the Art of Forgetting

Some archaeological remains are linked to stories about past events. The remains of St William's Chapel date back to the twelfth century.[69] The chapel was originally dedicated to St Catherine and had a cell of monks residing nearby who were linked to Norwich Priory. It was rededicated in 1168 to St William of Norwich. William was a 12-year-old boy who was an apprentice tanner. When his body was found on Mousehold Heath, it was claimed he had been murdered by Jews in a ritual killing, though there was insufficient evidence to convict them. William's body was initially buried where it was found but was later moved to the monks' cemetery. The chapel became a place of pilgrimage,[70] and by the nineteenth century had become a setting for tales of ghosts and witches, such as the Lady o'the Heath, the Crone o'Dussindale, and the black monk o'St. Williams.[71]

Another historic event associated with the Heath that has been mythologised is Kett's Rebellion of 1549. What is particularly interesting in this case is the way in which the memory of that event has changed with the prevailing spirit of the times. Mousehold Heath had been a location for political gatherings before the mid-sixteenth century, for instance, during the Peasants' Revolt of 1381. Robert Kett and his associates were perhaps aware of that history when they set up camp in 1549. They camped on Mousehold for seven weeks and it was from there that they issued their list of demands.[72] At the height of the protests, as many as 16,000–20,000 people may have been gathered on the Heath. The precise location of the camp is unknown, but it may have been situated near Thorpe Wood. Treswell's 1589 map includes the location of "The Oke of Reformation so called by Kett and rebels". The protesters were involved in two battles in August 1549. On 1 August, the Marquis of Northampton's army encountered the protestors at Bishopgate to the west of Thorpe Wood. The rebels were successful, and Norwich came under rebel control. The second battle, between 24 and 27 August, saw the protestors forced back from the city to the Heath by the Duke of Warwick's army. The protestors continued to attack the city from Mousehold, but then relocated to Dussindale where they were finally defeated. It has been suggested that Dussindale was in the area now known

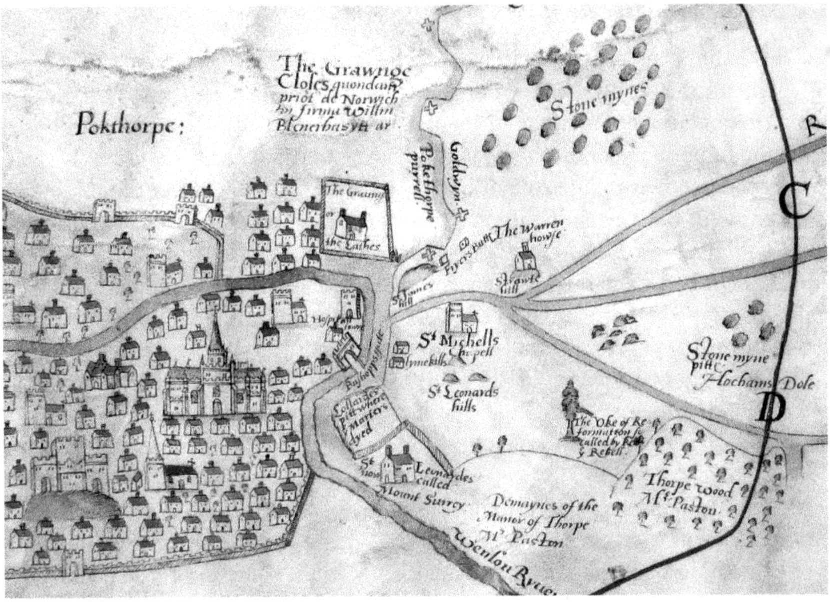

Figure 3.4 Ralph Treswell Map (1589).

as Long Valley. It has certainly been used for military activity, with 24 lead balls, six lead shot, and 15 musket balls found there,[73] but some of these are of a later date and there is no firm evidence that it was the site of the battle. Having been tried in London, Kett was brought back to Norwich to be executed and his body was hung from Norwich Castle.

On the 400th anniversary of Kett's Rebellion, Alderman Fred Herderson commissioned a stone plaque for the Castle walls which read: "in reparation and honour to a notable and courageous leader in the long struggle of the common people to escape from a servile life into the freedom of just conditions".[74] These sentiments have remained dominant since,[75] but as Nicola Whyte has demonstrated, this interpretation is out of line with earlier views. The Treswell map is typical of those views in its labelling of Kett as a rebel. Moreover, as Whyte notes, it conflated Kett's rebellion with Catholicism and with the expulsion of the Jews in the twelfth century, thereby presenting a narrative of local elites ridding the area of a variety of disruptive forces that threatened the social fabric. Similarly in Alexander Neville's memoirs, Holinshead's chronicles and Blomefeld's account, Kett is presented as "a villain and traitor, the usurper of the natural social order".[76]

While Kett's rebellion is the best remembered example of the Heath being used for political gatherings, it is not the only one. Journeymen weavers took to the Heath in the 1820s to highlight economic grievances and Chartist meetings were held there in the 1840s.[77]

Commemoration and the Recreation of the Historic Landscape

Mousehold has a history of rebellion, industry and community, the remains of which are visible in the topography of the landscape. The mines, wells and brickworks, Kett's Heights and the markers of the ancient chapel of William in the Woods all hint at the importance of the Heath across the centuries. The oral histories revealed various perspectives on the role of history in preservation and in encouraging engagement with the space. Should the history of common land be clear and present on the land itself through information boards, and walking tours and QR codes, or is that unnecessary? All the interviewees discussed the role of commemoration and public history, and were asked to consider whether an awareness of history might help in the effort to preserve and save common land from development.

Oral History Interviewee 1 (Norwich) explained that working for the Ramblers Association had demonstrated that the history of a landscape was not equally as significant to everyone who visits:

> It's exactly the same with a footpath out in the countryside. It may be once that there was a tremendous battle over getting it recognised as a right of way or something or getting problems dealt with, but if someone goes and walks it, they just go and walk it and enjoy it. They don't know about the history, they don't want to know, and so on.[78]

Oral History Interviewee 2, a retired history teacher, pointed to a couple of information boards up on the heath that indicate or hint at the history of the landscape: "'Here was a brick kiln' and, 'This was a particular chapel', and so on and so forth. I'm not sure how much more you can … do than that".[79]

On Mousehold Heath, the issue of commemoration of historical events appears less important than the debate around recreating historical landscapes. Efforts to "re-heath" the area were very popular among a section of interviewees, and it was seen as an attempt to revitalise the natural state of the soil and the land itself, a chance to return the land to the blasted Heath suggested by the paintings of the Norwich School in the eighteenth and nineteenth centuries.

Oral History Interviewee 3, a public health academic, argued that the modern needs of the city might supersede the benefits of reinstating the historical landscape. She argued that the pressure to protect habitats was legitimate, but that, overall, a woodland was a more valuable asset to public health than the heath:

> And so, I have argued with the ranger, when they do all these maintaining of the heath, cutting down trees … And now we know, it's come out this year … we need a trillion trees planted in the UK in order to sequester carbon. So where are they going to come? Well, we could plant them on city streets, but it just seems obvious.[80]

She concluded that there was a "slice of time" approach taken by some environmentalists, whereby conservationists would select an historical period and recreate that, in this case the heathland. This approach, she argued, did not necessarily reflect either the natural progression of the land or the most historically accurate way to approach conservation: "[Trees are] trying to come back all on their own on Mousehold, and it's not just a single tree being planted, but there's a whole ecosystem, a woodland ecosystem trying to regenerate itself".[81]

Yet, heathland is seen by some experts as the natural use for the acidic soil and brings with it environmental benefits of its own, even if these are not as obvious as woodland. Oral history Interviewee 4, an environmental officer for Norwich council, defined the Heath as a "used site" that has gone back to nature now that the gravel pits and bricks works are no longer functioning. He stated that "what people see as the woodland and the remnants of the heath at the moment are very recent, even though they are, to a certain extent, preserving what it would have looked like two or three hundred years ago".[82] He argued that tree clearance preserves the national and beneficial heathland:

> If anything, you would want to take more trees off of Mousehold to make it more open and more heath-like. But obviously you're balancing that with the fact that most people recognise Mousehold as being a site with lots of trees these days. And there's a balance there to be had and a conversation to be had with the public around what that site is or should be and how they want to use it.[83]

This perspective emphasises how invested local communities are in the development of Mousehold Heath. S/he explained that the idea of Mousehold is a misconception, but it reveals how residents of Norwich are attached to the image of Mousehold they have in living memory. S/he stated that when there are "any local stories about things that might be happening to the heath, whether it's cutting the trees down to improve the heathland, you'll suddenly get a flurry of local media stories". These were usually "erroneous", claiming that the trees had "been there for hundreds of years". Oral History Interviewee 4 clarified:

> Well, no they haven't, because if you look at a photo from fifty years ago and there aren't that many trees there and things like that. So, people have a kind of … there's a slightly misremembered history sometimes, I think.[84]

Oral History Interviewee 5, a Mousehold ranger, explained how he navigates this attachment through collaboration:

> My work was very much to, you know, engage with the local community and, in many ways, and to really, you know, try and get the message out there and try and show the benefits of the heathlands, such as the cultural benefits, so, you know, generations of Norfolk people had worked on the heathland, and also the beauty of the heather, the aesthetic value, and also the biodiversity. So, you know, if you go up there in summer when the heather is in bloom, there's thousands of bees, butterflies, the bare areas of the heath that

> would have been disturbed in the past, there's important species of solitary bee and wasp. So, there's lots of benefits. And this was under threat of being lost all together.[85]

This holistic approach marries the cultural history of Norwich with the natural history of the Heath and new environmental concerns like habitat preservation. Producing a detailed and publicly available management plan allows for transparency, while volunteering opportunities engage local communities with the work and the history and theory behind preserving diverse landscapes that do not appear as lush, green and environmentally conscious as woodland.

Contemporary Mousehold Heath: Environmental Governance and Management

Today, Mousehold Heath is an actively managed open space, with an ongoing programme of environmental improvement work and active community and educational engagement overseen by the Mousehold Conservators, two full-time wardens and a team of volunteers, voluntary groups and site wardens. The arrangements for the management of the Heath made by the 1884 Scheme Confirmation Act have continued to the modern day, albeit in modified form. Under the City of Norwich Act 1984, the management of the Heath remains the responsibility of the Mousehold Heath Conservators, an independent body which is, for all practical purposes, treated as a committee of the City Council funded by an annual precept and meeting four times a year. Under the 1984 Act, there are 12 Conservators, of whom nine are appointed by the Council (at least seven of which must be city councillors), one appointed by the Mousehold Heath Defenders (a voluntary body), one by the Norwich Society and one appointed "by a body with an interest in the conservation of the environment of the city". The Conservators introduced the first Mousehold Heath Management plan in 2008 following an inclusive public consultation process. The draft plan was also commented upon prior to approval in amended form by Natural England and the Norfolk Wildlife Trust.[86] The management plan is a "living" document intended to give a dynamic oversight and strategy for management that can change as environmental conditions and the financial context change over time. The original plan was revised in 2013.

The current management plan covers a ten-year period from 2019, with the overall vision to safeguard the long-term future of Mousehold Heath "as a high quality and accessible natural area for residents and visitors to the city to enjoy" and "to manage the heath in a way that protects and enhances the valuable historic and natural features of the site and provides opportunities for all to be involved in looking after and learning about the site".[87] Its eight management objectives focus on ensuring the safety of the Heath for visitors, its cleanliness, improving the natural environment and biodiversity of the Heath, and safeguarding its historic landscape and archaeological features. There are also two objectives focussed on community engagement that reflect the active "engaged management" approach that the Conservators have adopted: providing "opportunities for local communities to be involved in all aspects of our work" and promoting Mousehold Heath "to increase awareness, knowledge, understanding and a sense of pride".

The contemporary Heath supports a mosaic of different habitats, including acid grassland, heathland, mature woodlands, specimen trees, scrub and also some aquatic environments. Of the contemporary land area of Mousehold Heath, however, only approximately ten hectares are surviving heathland. Since the end of World War Two, especially since 1975, there has been a marked decline in the open heathland habitat. This has been subject to encroachment by dwarf shrub and tree populations – principally birch and oak, with some rowan – and is today in poor conservation status.[88] The City Council is a member of the Norfolk Biodiversity partnership, which has identified lowland heath as a priority habitat for restoration. And the Green Infrastructure Strategy for Norwich, adopted in 2007, prioritises Mousehold Heath as a key historic component for heathland regeneration in the north-east of the city. The current management plan divides the Heath into eight compartments (A-I). Heathland recreation work has been undertaken in compartments A, B, C and D, including topsoil stripping, and the removal of trees, scrub and gorse to encourage heather regeneration. A pilot heathland regeneration project has been undertaken in compartment D, which includes a good area of dense and mature heather and gorse. Although there is encroachment by birch, where topsoil stripping has been undertaken, this has encouraged good heather regeneration.[89] In 2012, the Norwich City Council was awarded a Higher-Level Stewardship agreement by Natural England to support the restoration of sections of heathland across Mousehold. This provides an annual maintenance grant of £30,600 towards work to regenerate areas of heathland and acid grassland up to 2022.

Some of the Heath's biodiversity and geological features also receive formal protection under contemporary environmental legislation. The Heath is designated as a Local Nature Reserve and a County Wildlife Site. The southern section of the Heath around St. James' Pit is also notified as a geological Site of Special Scientific Interest[90] because of its importance as a key locality for the future taxonomic and evolutionary study of Upper Cretaceous mosasaurs, a dinosaur of which teeth and vertebrae have been found on the site. The designation imposes restrictions on activities that can be lawfully undertaken within the Sites of Special Scientific Interest (SSSI) without the permission of Natural England, such as dumping or spreading materials, erecting structures, extracting minerals or introducing (or changing) woodland or tree management on the site.[91]

The site of the former St William in the Wood Chapel is a scheduled ancient monument under the Ancient Monuments and Archaeological Areas Act 1979, which gives protection against disturbance or development on the site without the permission of Historic England.[92]

The oral histories of Mousehold suggest that it is now used predominantly for recreational activity. Interviewees spoke about sledging down the steep rivets carved by centuries of mining, walking dogs across the heathland and the woodland, their children's favourite trees and their own childhood favourites too. They spoke about seasonal celebrations like bonfires, guys, fireworks and sharing food with the community. Interviewees remembered the informal football matches that were impromptu and collaborative, as well as the league football and the nearby Gothic Social Club that sold cheap beer and hosted bands:

> I must admit, I miss the football now. Some of the pitches now for football aren't used much more now at all really. Very, very sparingly, really, to be honest with you. But it doesn't… but they're still there to be used for like walking your dog and things, really. So, there's always consistent things through spaces, really.[93]

This interviewee's conclusion here was an excellent summation of the physical importance of the commons. The space itself is adaptable, there is always a use for it and it is always valuable:

> I've seen people there the other day, they were doing boxing, so someone was teaching some boxing … there was football, running … it's a multipurpose place, I think, that a lot of commons don't really have. You know, so you can go for a picnic, you can go for a walk, you can take your dog, you can play football, you can play badminton.[94]

Mousehold Heath retains its popularity today. It is less of a place of political dissent and protest than a place of recreation and especially during the COVID 19 pandemic, a refuge from the city during lockdown. People still use it, and take from it, but what they take is less physical and more psychological. Interviewee 3, a public health academic, spoke about her research on the connection between mental health and green space. She discussed the ways in which public health bodies like the National Health Service (NHS) were, even during the pandemic, still pushing the importance of exercise, being outdoors and "environmental protection". She underlined that even with the pressures of COVID, the NHS announced that they have planned to go carbon neutral by 2040.[95] It is clearly a top priority:

> And there's such an obvious health link. Either because of the effects of extreme weather or just … the co-benefits. So, if you have, for example, active travel, you know, then you get health benefits and you get carbon reduction … So, I think in public health, so I would say a lot of the spirit behind urban commons is similar to the spirit behind public health. It just looks at the health outcomes … the things that urban commons maybe hasn't quantified in health terms.[96]

Defining urban commons in the modern context must take health and psychology into account. Like libraries, commons exist as space where there is little expectation to spend money, and combined with the suggested health benefits of exercising outdoors, they clearly have a vital role to play in the health and well-being of a city overall. Interviewee 3 emphasised the interaction between health and wealth inequality by stating that "for a population to be healthy" they need the job stability, housing and "a sense of community".[97] This community is created through

Mousehold's programme of events, volunteering schemes and its communal location. As demonstrated earlier, residents of Norwich have long relied on Mousehold. They still do. In a modern context, the newer threats of climate change, austerity and development force those managing urban commons to strike a balance between renewed interest in outdoor spaces and increased traffic, the maintenance of biodiversity, generating income and providing the services that encourage community involvement and commitment.

Present Voices, Future Directions

These contested positions play out powerfully in the image that the current users of Mousehold Heath have for the future of this green open space. Concerns over the respective health and biodiversity benefits of heath or woodland and the interrelation between future protection and increased engagement in numbers of people and modes of activity were raised in community visions for Mousehold Heath. Participatory research activities undertaken for this project included in situ and walking conversations with conservation volunteers and regular users; mapping workshops with participants at a men's mental health support group, school children, members of a youth club and with public users; and the co-creation of a performance-for-film with young adults at a youth engagement project.

The restoration of the heathland for the future was important to most participants, evident in conversations with conservation volunteers and in illustrations of purple heathers across participants' maps of the Heath in 2050. Where imaginings clearly protected the heathland, there were also visions of establishing "a thicks", an area of ancient woodland comparable to the one at Staverton, with an understanding that preservation of woodland into the distant future relied on land management decisions today.[98] Conflicting views on extending heathland in relation to woodland have been well aired in this chapter. Certainly, conservation volunteers attested to experiencing "a lot of flack in cutting down the trees", demonstrating how tensions manifest on the ground.[99] Added to this debate, however, must be a consideration of the volunteer labour required to enact the Conservators' plan. Restoring future heathland to a fixed point in the past aims for the extent of heather captured in a 1970s aerial photograph rather than old landscape paintings, but its achievement is still reliant on a significant contribution of volunteer labour hours.[100] One conservation volunteer, Penny, explained that their efforts to expose the heather were "a never-ending task" because the bracken "always grows back".[101] She concluded that pulling the bracken out by the roots did reduce regrowth; while her demonstration of root pulling by hand compared to the nearby sweeps of a brush cutter served to illustrate the labour-intensive nature of the task. Another volunteer, Stella, similarly identified the value of volunteers' manual labour:

> Will [the warden] always tells us how useful the work that we're doing is. And, yeah, it does make you feel good. Because obviously… nobody is going to be paid doing something like this. And it's one of the few jobs, I suppose now, that's sort of manual, isn't it. It's something that you can do quite

> easily, especially the more people you have, the easier it is. But they're never going to pay anyone, it wouldn't be feasible to conserve the heath in the way we do it, sort of manually pulling up bracken and stuff like that.[102]

Her reflection emphasised the reliance of the heathland's restoration on a supply of volunteer labour, especially to perform manual, and perhaps less environmentally invasive, conservation practices. Her observation that "they", the City Council, will never pay anyone to undertake the work is indicative of the constraints on funding green space management as local authorities struggle to meet increasing and competing demands for public services. Where volunteer schemes support the engagement of local people in the work of the Conservators, an over-reliance on volunteer labour could have implications for the future quality and quantity of land management.

Yet, Stella's comment also highlights the reciprocity of the volunteer contract. Her labour is exchanged for a currently scarce opportunity to undertake physical work and positive emotions enlivened in her fulfilment of conservation tasks. Another volunteer, Sian Rowlands, identified specific psychological and physical benefits of her participation:

Figure 3.5 Mousehold Heath Mouseketeers Volunteer Working Party carrying out conservation tasks, Mousehold Heath, Norwich, 2021.

Image: Siobhan O'Neill.

> I feel I'm doing something worthwhile, so that feeds into my self-esteem, my self-worth, stuff like that. I'm quite shy so it helps get me about of myself, helps me to meet people. And then obviously the physical benefits. If you're busy doing stuff you don't have time to be worried about the past, the future. It cuts down on the anxiety a bit. [103]

Engagement in bodily exercise, "being out in nature",[104] the acquisition of new skills and co-operative working were perceived by all volunteers to support physical health, social integration and well-being. A sense of community was connected to both the "social occasion"[105] of the working party and to an understanding of "giving something back"[106] to the broader community. Robert Stubbs, a retired NHS administrator, identified the benefits in terms of "getting out there and trying to be more sociable", alongside contributing towards improvements, such as a new cycle path that will "be there, probably not forever, but for quite a while".[107] Just as the notion of common land encourages access, so too acts of collective care-taking nurture feelings of individual pride and community ownership or as Robert Stubbs concluded:

> Stewardship is a better word for it. Working to improve the area for that particular time and for future generations.[108]

In this reciprocal encounter, then, the experiences of self-worth, health and well-being engendered through the "engaged management" approach may prove key to ensuring the ongoing commitment of volunteers upon which the future of Mousehold Heath's biodiversity and protection rely.

How future generations wish to engage with Mousehold Heath, however, do not always fall within current expectations of appropriate use. One such field of conflict is the leisure sport, BMX biking. As one respondent to a 2019 Norwich City Council questionnaire about the future of the Heath replied, "the Mousehold Conservators have allowed areas of the heath to become [a] bike track so clearly they are not committed to safeguarding it for the long-term future", to which the Mousehold Conservators replied, "The wardens take action to deter these activities".[109] The issue of safeguarding seems to reflect the intense physicality and risk-taking that characterises the sport in contrast to more normative leisure pastimes, such as walking, with implications for riders and walkers occupying the same space. It also refers to the protection of the landscape; bikers have undertaken wild builds in some areas, physically moulding the terrain to construct bike trails and jumps.

Where some public users have been vocal in their complaint of the bikers' use of the Heath, participants in this research were generally more open to it as a form of young people's engagement. One conservation volunteer, Robert Stubbs, commented, "as long as there isn't an excessive amount of damage", then Mousehold Heath "is such a large area, it can cater for most things". Another volunteer, Rosamund Chettleburgh, emphasised the benefits for teenagers:

> …and they weren't on a computer, they were doing things like cycle rides with their mates and using the heath. It's got to be a good thing for them.[110]

Asked to identify the value, she highlighted the alleviation of "mental health issues" and "anxiety",[111] which she attributed to increased academic pressures and life choices. Support for well-being was also identified by two BMX bikers, one a care worker and the other an engineering student, who explained how practicing BMX skills helped release the tensions of the day. The provision of a dedicated space to ride bikes it seemed could support community well-being for young people.

The understanding of BMX biking as a sport was raised by Norfolk-based geologist Tim Holt-Wilson:

> They should be allowed to get on with what they're doing now because actually it's a tremendous creative resource for spectacular gymnastics. And as we've seen in the recent Olympics, great things can be done on a BMX bike.[112]

In this reflection, he identified the athletics of BMX biking and its validation as a competitive sport; it first appeared in the Olympic programme in 2003. Certainly, observations of how the two bikers, mentioned above, practised a specific jump time and again to acquire the physical skill to perfect it demonstrated dedication indicative of a sport.[113] This notion of BMX biking as a sport helps to contextualise the transition from riders simply taking advantage of Mousehold's undulating

Figure 3.6 BMX dirt track with bike jump insert, St James' Pit, Norwich, 2021.

Image: Siobhan O'Neill.

terrain to the construction of various trails to accommodate different skill levels. In a description of the wild build in St James' Pit, Tim Holt-Wilson emphasised the "creative, collective genius" in the way the bikers "have terraformed the wood into a BMX Track",[114] concluding with an analogy:

> I mean like you look at termite mounds and you feel this tremendous sense of awe because of what they've managed to do.[115]

As a geologist, Tim Holt-Wilson appreciated the ingenuity of the young adults who "mounded" and "shaped" the soil into trails with "steeply cambered embankments that you can whizz around at great speed...and things you can jump over".[116] Moreover, he recognised the collective act of them co-operatively working to generate a shared recreational asset. Having worked with the Conservators to create the Mousehold Heath Geological Trail, Tim Holt-Wilson understood the area is a designated geological Site of Special Scientific Interest, but argued, "it's all in the spoil of the old pit, so it's not geologically sensitive".[117] His assertion therefore that the bikers "should be allowed to get on with what they're doing" referred both to bike riding and dirt track building, an assertion he "would argue in front of any tribunal".[118] These comments highlighted the intensity of feeling evoked in support of, as well as against, young bikers creating and managing a leisure facility for themselves in the Heath.

In actuality, St James' Pit lies outside the boundary of "legal" Mousehold, and the wild build makes use of an adjacent piece of open public space besides Heathgate.[119] In the public imagination, the boundary of the Heath seems unclear and abutting green spaces are not always understood to fall outside of the jurisdiction of its bye-laws. Considered as a sporting facility, then, parallels might be drawn between the BMX bike track and the municipal golf course, similarly housed in land adjacent to the Heath. Concerns over environmental impact may be weighed in different ways at both leisure resources; however, the issue of public liability is particularly pertinent for the community-led, rather than civic, initiative of the BMX track. In our engagement with the BMX community at St James' Pit, bikers were open to exploring the possibility of constituting themselves as a Club to seek public liability insurance in order to safeguard the future of their sporting activity and dirt track through permission from the local authority as landowners. There are precedents for young people working with local authorities to redevelop dirt tracks;[120] however, "development" tends to formalise building practices in a way that seems to contradict the sense of collective, do-it-yourself activity here, which is characteristic of commons.

An ambition that arose across all interventions with research participants was to increase learning and engagement opportunities on Mousehold Heath. In interviews and workshops, participants expressed a desire to enhance education on the history, biodiversity and land management of the Heath. Volunteers, who acknowledge the learning afforded in their conservation work – practical horticultural skills, flora and fauna species identification, awareness of broader debates such as rewilding and climate change – were keen to extend similar opportunities to the wider local public, in particular young people. One volunteer, Rosamund Chettleburgh, suggested:

> I think we have to pick a pathway through to giving youngsters an opportunity to come and do something like this, maybe on a Sunday afternoon… Maybe somebody who's a trusted volunteer or whatever, maybe once a month in the summer to do something… I've no idea what age group, 12 to 18 something like that.[121]

Her recommendation that a "trusted volunteer" might lead the youth engagement is reflective of volunteers' understanding of restrictions on the time wardens can allocate to public engagement. Another volunteer, Robert Stubbes, explained, "it's that balance between actually doing this work and spending your time with groups coming up here".[122] The volunteer-led proposal, however, could be complicated by "how things are at the moment with healthy and safety and risk assessments",[123] as attested by volunteer and previous scout leader Paul Jerman. It is in part due to such issues of safety and liability that the scout group no longer comes out to do "night exercises" or forest skills in the Heath.[124]

Where the Conservator's management plan aims to increase public knowledge and understanding, the warden's role focuses on land management, leaving limited capacity to support an educational offer. At the time of this research, one independent school involved pupils in conservation work, [125] facilitated largely through the efforts of assistant head for enrichment and geography teacher, Owain Hall, who approached the warden with the proposal.[126] By programming the weekly participation of six Year 12 pupils to take place concurrently with the Mouseketeer

Figure 3.7 Siobhan O'Neill. In-*Common Sites*. Performance for film co-created with the Sprowston Youth Engagement Project, Mousehold Heath, Norwich, 2021.

Image: Annis Joslin (film still).

Working Parties, both groups could be supported by the warden. Employing this co-ordination enabled young people's engagement, though it would prove difficult to replicate for larger groups. Moreover, as an enrichment activity, the intention of pupils' participation was more to instil values than for education per se, as Owain Hall explained, for pupils to recognise that "it's actually really good and important to do things that aren't for you" and "to open their eyes to volunteering as they grow into adults".[127] As a useful model, however, it prompted volunteers to call for similar opportunities for state schools.

A Forest School initiative for a local state primary school delivered by freelance educator Nicola had been running in Mousehold, but it was discontinued during the COVID-19 pandemic. Nicola explained:

> I kind of thought, 'Oh, will I still be there?' and then in lockdown they straight away were like, 'No, you're not allowed to work here in lockdown' whereas my two schools that I'm still at that aren't academies with strong independent heads are like, 'Yeah, no, you can carry on working. We definitely want outdoor learning.'[128]

Though pandemic restrictions were the catalyst, Nicola attributed the closure more to the change in status of the school to an academy and the implications she perceived this had on finances, particularly the availability of funds for additional activities on pg. 54. Neither of her other schools use the Heath as they are not situated in the neighbourhood, and the issue of transporting pupils for short sessions is prescriptive for state-funded schools. In both examples, dedicated educators initiated and enabled school engagement, which the warden responded to with support. As such, educational initiatives tend to be ad hoc and are reliant on individual community champions, which mean provision is variable and precarious.

The aspiration to embed the provision of educational and engagement activity into the management of the Heath was forcefully articulated by participants in future mapping workshops with Pit Stop, a community organisation supporting men's well-being. A map designed with group facilitator Colin Howey incorporated:

> An outdoor education lodge, (and) performance space where you could do traditional crafts, forest schools, all sorts.[129]

In a discussion around the map, Colin Howey went on to indicate the types of activity he would like to see supported by the hub:

> well-being walking, nature walking, fitness, outdoor exercise, all sorts of stuff that could be activated in that space and inspired by that space.[130]

Discussions within the group recognised that walking and fitness activities go on informally; however, the notion of a dedicated education centre to activate engagement received strong support. One participant commented, "it's crying out for it. A lot of places round the country have got them".[131] In another map, drawn by participant Alex Helm, the recommendation for infrastructure extended to four hubs – a central information centre and three satellites to encourage various forms of engagement in different locations on the Heath:

> One of the hubs would be for forest schools and activities to teach children, early learning, teach them the importance of conservation. And they can get involved in conserving, or landscaping, or whatever kind of activities they want. One hub specifically for conservation, so that's adults, and that's ways they can improve and maintain… A hub for activities, with sports, performance and utilising the bandstand, for that as well. So, there's four central hubs. And probably a café at the main central hub.[132]

Alex Helm's plan encompassed visions expressed by many of the workshop participants. Again, the encouragement of children and young people to connect and learn about nature was significant. So too the involvement of adults in conservation work. In addition, creative engagement was highlighted in proposals for craft practices, the revitalisation of the bandstand and a performance space. The latter led to reflections on the community theatre production, 1549: The Story of Kett's Rebellion by Common Lot Theatre,[133] performed on the Heath in 2016, which had also been popular in conservation volunteers' recollections. In a walking conversation with the production's director and co-writer, Simon Floyd, he also commented on the educational value of performance, "giving people a sense of the history of a place", alongside the offer of multiple possibilities for engagement:

> There's a lot of participation possible because you've got a lot of singing, a lot of rebels, there's a lot of main characters. It's ideal, it's an ideal vehicle for community involvement, participation.[134]

In addition, the desire for a café was another feature that echoed across all participant groups, who identified the closure of a public café at the prison as a significant loss, specifically as a social space to gather pre- and post-venturing onto the Heath.

Within discussions on the various proposals, issues were raised specifically around how provisions could be financed and the restrictions placed on built infrastructure under the terms of Mousehold Heath covenant. Though no concrete solutions were defined, the thinking through of these issues demonstrated a clear appetite to maximise public engagement and use of the Heath as a space for learning, health and well-being for all generations.

Notes

1 Norfolk Record Office, Norwich, MS 4392, Judgement delivered on the case of the City of Norwich versus Browne and others to establish the title of the Corporation of Norwich to Mousehold Heath, 7 June 1881.

2 Andrew Macnair and Tom Williamson, *William Faden and Norfolk's Eighteenth Century Landscape* (Oxford: Oxbow, 2010), p. 102. By contrast, as the work of Sara Birtles

has shown, many of the Norfolk commons that were registered under the Commons Registration Act 1965 were not historic commons. S. Birtles, 'The Impact of Commons Registration: A Norfolk Study', *Landscape History* 20: 83–97.

3 Elizabeth Griffiths and Hassell Smith, *'Buxom to the Mayor': A History of the Norwich Freemen and the Town Close Estate* (Norwich: East Anglian Studies, 1987), pp. 10–12; Oliver Rackham, *'Houses of Benedictine monks: The Cathedral Priory of the Holy Trinity, Norwich'*, in *A History of the County of Norfolk: Volume 2*, ed. William Page (London: Victoria County History, 1906), pp. 317–328.

4 The present and future uses of the Heath will be considered in more detail in Chapter 9.

5 Norfolk Record Office, Norwich, MS 4392, Judgement delivered on the case of the City of Norwich versus Browne and others to establish the title of the Corporation of Norwich to Mousehold Heath, 7 June 1881.

6 For example, the Monastic and Cathedral accounts for 1532–1534 discuss the digging of sand: account of Robert Canon. Prior of the Office of Master Cellarer "of the digging of sand there, nothing. And of the sheep's pasture there this year 33s 4d", and for 1534–1535, it is noted that "of the digging of sand there this year nothing, because without a farmer". Norfolk Record Office, Norwich. Supplementary papers submitted, Norwich Corporation versus Browne and Others 1881, pp. 8–9.

7 In the seventeenth century, the cathedral authorities certainly seem to have managed the digging of sand closely: for example, the Pockthorpe Manor Court Leet Book records that on 6 November 1670, one Augustine Blennerhasset was fined 5 shillings for unlawfully digging "sand pitts" on the demesne lands and was ordered to fill them up within one month under penalty of forfeiture of 20 shillings. Supplementary papers submitted, Norwich Corporation v Browne and others 1881, p. 20, Norfolk RO.

8 For an excellent account of the local political tensions, political manoeuvring and the place of the Mousehold Heath case in the wider politics of the Victorian Parks movement, see Neil Macmaster, "The Battle for Mousehold Heath 1857–1884: Popular Politics and the Victorian Public Park" (1992), 127 *Past and Present* 117–154.

9 Public Health Act 1875, s. 164.

10 Indeed, the Dean and Chapter of Norwich Cathedral had themselves purchased gravel from them during the 1870s. See discussion in the *Norfolk Chronicle* for Saturday, 28 May 1881. The "easy going nature and laxity" of the Dean and Chapter was reported as having encouraged the Pockthorpians to build "clamps" and start the manufacture of bricks.

11 There are detailed accounts in the *Norfolk Chronicle* for Saturday, 28 May 1881, 18 June 1881 and 2 July 1881.

12 The 1881 Chancery proceedings are reported in some detail in the *Norfolk Chronicle* for Saturday, 6 August 1881.

13 Or *profit a prendre*, an incorporeal hereditament entitling the holder to take the natural produce from someone else's land.

14 Above, note 8. The 1875 Act provided that land could be transferred to be held as pleasure grounds "subject .... to rights of common, easements and other rights".

15 The Mayor and Corporation of Norwich v Brown and others (1883), 48 Law Times, 725

16 The Mayor and Corporation of Norwich v Brown and others (1883), 48 Law Times, 725 at 900.

17 Formerly, the Inclosure Commissioners.

18 Mousehold Heath Scheme Confirmation Act 1884, recitals.

19 See Christopher Rodgers, "Towards a Taxonomy for Common and Public Property", (2019) 78 *Cambridge Law Journal* 124–147.

20 Emphasis added.

21 Tom Williamson and Andrew MacNair, *William Faden and Norfolk's 18th Century Landscape* (Oxford: Windgather Press, 2010), p. 100.

22 Ibid.

23 Oliver Rackham, *The History of the Countryside: The Classic History of Britain's Landscape, Flora and Fauna* (London: Wiedenfeld & Nicholson, 1986), p. 301; Norfolk Record Office, Norwich, C/SCEI/16. For a grant of liberty to enclose 2 acres of Mousehold Heath and Common or Sugar Loaf Hill, 1775.
24 Norfolk Record Office, Norwich, COL 9/74. Communication from Mr John Wagstaffe upon the inclosure of Mousehold Heath, 31 August 1792.
25 Ibid.
26 Ibid.
27 Geoffrey Goreham, *Mousehold: A Short History* (Norwich: Mousehold Heath Defenders, 1973).
28 Unidentifed author, 'Tales of Mousehold Heath', in *The Norwich Magazine* (Norwich: Josiah Fletcher, 1835).
29 Ibid.
30 Rackham, *The History of the Countryside*, pp. 299–302.
31 Norfolk Record Office, Norwich, MS 4392, Judgement delivered on the case of the City of Norwich versus Browne and others to establish the title of the Corporation of Norwich to Mousehold Heath, 7 June 1881.
32 Ibid.
33 Norfolk Record Office, Norwich: DCN120/2LL.14. Memorandum; DCN 120/2LL.13. Correspondence between the Guildhall and the Dean and Chapter of Norwich, 23 January 1855.
34 Ibid.
35 Norfolk Record Office, Norwich: DCN 120/2KK/2. Letter marked private from George Pellaw (on behalf of the Canons of Norwich Cathedral) to 'the Rev'd', 16 November 1864.
36 Norfolk Record Office, Norwich: DCN 120/2LL/63. Abstract of the Canons reflections.
37 Ibid.
38 Ibid.
39 Ibid.
40 *City of Norwich Act, 1867. Draft Scheme with Respect to Mousehold Heath*, within judgement delivered on the case of the City of Norwich vs Browne.
41 Norfolk Record Office, Norwich: N/TC 52/10. Unemployed Mousehold Heath.
42 Norfolk Heritage Centre: 942.615. *A Unified Mousehold.*
43 *Norfolk Chronicle* and *Norwich Gazette*, 8 December 1838, 20 July 1839.
44 *Morning Chronicle*, 16 November 1824.
45 *Norfolk Chronicle* and *Norwich Gazette*, 23 June 1838.
46 *Bury and Norwich Post*, 4 July 1838.
47 Norfolk Record Office: MC 17/122, 543X8, 'Letter from Rosa Howes inviting her to tea on Mousehold Heath, Norwich'.
48 Norfolk Record Office: MC 2487/74, 977X3, 'Letter from John Sell Cotman to Dawson Turner'.
49 Andrew W. Moore, *The Norwich School of Artists* (Norwich: Norfolk Museums Service, 1985).
50 Cited in Nicola Whyte, 'Remembering Mousehold Heath', in *Remembering Protest in Britain since 1500: Memory, Materiality and the Landscape*, ed. Carl J. Griffin and Briony McDonagh (Cham: Springer, 2018), p. 41.
51 For the bye-law powers see generally the Mousehold Heath Scheme Confirmation Act 1884, Sched. 1 para 17.
52 Whyte, 'Remembering Mousehold Heath', p. 42.
53 Norfolk Historic Environment Record (NHER): 58089, *Norfolk Heritage Explorer,* Norfolk County Council, https://www.heritage.norfolk.gov.uk/
54 NHER 26550, 26561.
55 Norfolk Record Office: DCN 120/2LL/15, 'Copy of two Chapter orders reflecting Pockthorpe Waste dated 1801 and 1828'.

56 NHER: 2656.
57 NHER: 12415.
58 NHER: 53589.
59 NHER: 631.
60 *Norfolk Chronicle,* 24 January 1835.
61 *Norfolk Chronicle,* 3 June 1826.
62 Norfolk Record Office: NCR 6a/42/30, 'Inquisition'.
63 Norfolk Record Office: NCR 6a/8, 'Inquisition'.
64 Norfolk Record Office: NCR 6a/8; NCR 6a/29, 'Inquisitions'.
65 Norfolk Record Office: NCR 6a/8.
66 *Bury and Norwich Post*, 24 July 1822.
67 NHER: 670, 8125.
68 Rose, *Mousehold Heath: A History*.
69 Geoffrey Goreham, *The Chapel on Mousehold Heath* (Norwich: Norfolk Heritage Centre, 1969).
70 NHER: 824.
71 *Tales of Mousehold Health* (Norwich: Josiah Fletcher, 1835), pp. 147–180.
72 Anthony Fletcher and Diarmaid MacCulloch, *Tudor Rebellions: Revised 5th Edition* (London: Routledge, 2014), pp. 156–159.
73 NHER: 53875; Norfolk Museums Collection, Accession Number: NWHCM 1973.296, 1974.546.1, https://www.museums.norfolk.gov.uk/collections.
74 Whyte, 'Remembering Mousehold Heath', p. 28.
75 Whyte, 'Remembering Mousehold Heath', p. 29.
76 Ibid.
77 'Chartist Meeting', *Norfolk Chronicle and Norwich Gazette*, 2 May 1840, p. 2.
78 Oral History Interview (Norwich) 1 ("OHI Norwich" 1).
79 OHI Norwich 2.
80 OHI Norwich 3.
81 OHI Norwich 3.
82 OHI Norwich 4.
83 OHI Norwich 4.
84 OHI Norwich 4.
85 OHI Norwich 5.
86 Semi structured interview 16 April 2020.
87 Mousehold Heath Management Plan 2019, Appendix 1, para 4.
88 See Mousehold Heath Management Plan 2019 at pp. 19–20.
89 See Mousehold Heath Management Plan 2019 at pp. 18, 19 and Map 4 for a detailed account of the works undertaken in each compartment of the Heath.
90 Wildlife and Countryside Act 1981, s. 28.
91 St James' Pit SSSI, Operations Likely to Damage the Special Interest, OLD 2000158, refs. 7, 12, 20, 23. Wildlife and Countryside Act 1981, s. 28E.
92 Ancient Monuments and Archaeological Areas Act 1979, listing 1003172.
93 OHI Norwich 6
94 OHI Norwich 7
95 OHI Norwich 3L
96 OHI Norwich 3
97 OHI Norwich 3
98 Walking Conversation with Colin Howey on 24 August 2021 with reference to Staverton Park and The Thicks in Suffolk.
99 Walking Conversation with Paul Lynch on 23 November 2021.
100 Volunteer schemes on Mousehold Heath comprise the Mousehold Heath Mouseketeers (two fortnightly working parties): a term-time weekly group of pupils from an independent school and intermittent support from TVC.

101 Walking Conversation with Penny on 17 August 2021.
102 Walking Conversation with Stella on 17 August 2021.
103 Walking Conversation with Sian Rowlands on 13 September 2021.
104 Walking Conversation with Stella on 17 August 2021.
105 Walking Conversation with Penny on 17 August 2021.
106 Walking Conversation with Rosamund Chettleburgh on 17 August 2021.
107 Walking Conversation with Robert Stubbs on 17 August 2021.
108 Walking Conversation with Robert Stubbs on 17 August 2021.
109 Norwich City Council. (15 March 2019) *Mousehold Heath Conservators, Meeting Minutes Appendix 2: Mousehold Heath management plan 2019–2028: Consultation summary*. shorturl.at/ekqrP.
110 Walking Conversation with Rosamund Chettleburgh on 17 August 2021.
111 Ibid.
112 Walking Conversation with Tim Holt-Wilson on 17 August 2021.
113 Observation of BMX bikers, St. James' Pit, 8 December 2021.
114 Walking Conversation with Tim Holt-Wilson on 17 August 2021.
115 Ibid.
116 Ibid.
117 Ibid.
118 Ibid.
119 See Figure 3.3. Map depicting the erosion of Mousehold Heath, on page 54.
120 See Anna Malina, *Proposal for BMX Track on Silloth Green* (Silloth Town Council, 2012). Available at: https://cumbria.citizenspace.com/other-public-sector/silloth_bmx_track/results/compressed_final_hlf_bmx-report-july-2012.pdf
121 Walking Conversation with Rosamund Chettleburgh on 17 August 2021.
122 Walking Conversation with Robert Stubbs on 17 August 2021.
123 Walking Conversation with Paul Jerman on 21 November 2021.
124 Ibid.
125 Norwich School for Girls offers six pupils each term to undertake 1.5 hours a week of conservation work, and whole year group, Year 9, to participate annually in a one-day conservation event.
126 Walking Conversation with Owain Hall on 13 September 2021.
127 Ibid.
128 Walking Conversation with Nicola on 14 November 2021.
129 Colin Howey: Future Mapping Workshop with Pit Stop, 22 November 2021.
130 Ibid.
131 Participant: Future Mapping Workshop with Pit Stop, 22 November 2021.
132 Alex Helm: Future Mapping Workshop with Pit Stop. 22 November 2021.
133 Floyd Simon and Minns Karl, *The Story of Kett's Rebellion.* Directed by Simon Floyd, A Common Lot, 2016, Mousehold Heath, Norwich.
134 Walking conversation with Simon Floyd on 22 November 2021.

4 Clifton and Durdham Downs, Bristol

Introduction

Location

Clifton and Durdham Downs (known collectively as "the Downs") consist of 442 acres of meadow and woodland to the north-west of Bristol city centre that is jointly owned by Bristol City Council and the Society of Merchant Venturers. Today, the Downs are bounded by the Bristol suburbs of Clifton, Redland and Stoke Bishop and by the Avon Gorge. Historically, they consisted of two adjoining commons located within the manors of Clifton and Henbury, forming a "wrangle common" over which the commonable cattle of both manors could range as a form of trespass.[1] Open arable fields and meadows separated the villages of Clifton and Henbury from nearby Bristol, but commoners had access to a series of turf tracks connecting the Downs to the urban settlement. The urbanisation of the area around the Downs since the eighteenth century has been crucial in determining the shape and character of this historic common.

Topography/Features

The Downs sit on Tickenham Ridge, a carboniferous limestone ridge that can be observed within the cliff face of St Vincent's Rock. The shallow calcareous soils covering the ridge are topped by open grassland, suitable for grazing and with easy opportunities for the shallow extraction of minerals. In the nineteenth century, dolomite and oolite were extracted for use as building stone and for the production of cement, creating a direct connection between the landscape geology of the Downs and the built environment of Bristol. Owing to the distinctive geological history, the Downs have long been a site for fossil collecting. In 1842, workmen found a large number of remains in a fissure in one of the quarries on Durdham Down, including bear, hyena and even hippopotamus, rhinoceros and elephant bones.[2]

While the Downs predominantly consists of open grassland, incorporating some historic species, areas of ancient and semi-natural woodland have been preserved. Substantial tree clearance began during the Neolithic period and continued throughout the Bronze Age. The open grassland was then maintained by sheep

DOI: 10.4324/9781003204558-5

Figure 4.1 Map of Clifton and Durdham Downs.

grazing, which prevented the growth of self-sown seedlings. In the 1870s, the Downs Committee began a planting scheme that included the introduction of alien species such as Austrian Pines and Huntingdon Elms. This impacted on the native species, but in the twenty-first century, the clearance of some imported species has aided the recovery of rare plants within the Gully. The Huntingdon Elms that were planted along Ladies Mile were affected by Dutch Elm disease and were replaced by Limes in 1980.

Narrative

Archaeological assessment has established that there has been continuous human interaction with the landscape of the Downs since the Neolithic period.[3] The Iron Age hillfort Clifton Down camp is a scheduled ancient monument that covers approximately 2.6 hectares.[4] The camp was one of three forts that lay on either side of the Avon Gorge. Parts of the earthworks were levelled in the eighteenth century and large sections were quarried during the nineteenth century to provide material for new roads.[5] There are also low banks of turf north of Clifton Down Road that may be the remains of Iron Age or more likely Roman field boundaries.[6] The remains of the Roman road that ran between Portus Abonae (now Sea Mills) and Aqua Sulis (Bath) are still visible in the form of a grass-covered bank that runs east to west across Durdham Down. Moreover, the discovery of Roman coins within Clifton Down hillfort as well as of a finger ring featuring a portrait of the Emperor Nero suggest that the camp may have been reused under the Romans.[7]

The earliest documentary evidence for grazing rights pertaining to the Downs dates to an Anglo-Saxon charter of 883CE. The charter detailed the administrative boundary between the manors of Clifton and Henbury, which ran from the bottom of *Eowcumbe* (Walcombe Slade) to a site close to the present water tower.[8] This boundary is preserved today by seven replica merestones that run across the Downs.

Clifton Manor was held by Lewin the King's reeve from 1042, passing to Robert FitzRalph in 1086 and to William de Clifton in 1150. In the fifteenth century, it was split into Greater and Smaller Clifton. The former continued to be held in private ownership, but the latter was granted to the Abbey of St Augustine and later to the Dean and Cannons of the College at Westbury on Trym, returning to private hands at the dissolution of the monasteries. In the seventeenth century, first Greater Clifton and then its smaller neighbour were bought by the Society of Merchant Venturers, bringing Clifton Down under its jurisdiction. By the seventeenth century, Henbury Manor (including Durdham Down) was under the authority of three or four landowners who controlled land use through fines and leases.

From the mid-eighteenth century, the area, which had previously been largely untouched by urban expansion, was impacted by the development of Clifton as a fashionable residential area and Hotwells as a spa resort. This brought increased building along the edges of the Downs, which also resulted in some losses to the green space through encroachment and enclosure. By the mid-nineteenth century, grazing was much less common, but quarrying continued, and recreational activity – encouraged by the eighteenth-century developments – became increasingly dominant.

During the 1850s, concerns regarding damage to the Downs by quarrying and vehicular encroachments were increasingly raised. In 1856, the Society of Merchant Venturers promised "to maintain the free and uninterrupted use of [Clifton] Downs". The following year, the Council sought to protect Durdham Down by purchasing the two tenements at Stoke Bishop for £450 that enjoyed common rights, and in spring 1858, sheep stamped "CB" were put out to graze on the Downs to maintain the rights of pasturage.[9] In 1859, the Council established the Downs Encroachment Committee and entered into negotiations with the Lords of the Manor of Henbury and Westbury. These negotiations came to fruition in the Clifton and Durdham Downs (Bristol) Act (1861), secured through collaboration between the City Council and the Merchant Venturers. The Act brought Durdham Down under Council ownership so that both commons could be protected from development. The management of the area was to be overseen by a Downs Committee, which included among its membership seven city councillors and seven Merchant Venturers and was chaired by the Lord Mayor of Bristol. The Committee quickly appointed a Downs Ranger to manage the space.

Figure 4.2 "Survey of the Manor of Clifton" by Jacob De Wilstar (1746).

Changing Space Over Time

The "Survey of the Manor of Clifton", produced by Jacob De Wilstar on behalf of the Society of Merchant Venturers in 1746, gives a sense of the extent of the Downs at that time.[10] De Wilstar claimed that the common land within the parish of Clifton extended to 370 acres. This figure was subsequently questioned by John Latimer, who pointed out that according to the 1861 Act, the area of Clifton Down was 230 acres.[11] Latimer argued that the earlier measurement was an exaggeration, but it is possible that De Wilstar's figure referred not just to Clifton Down, but to all the wastes, roads and lanes within the parish. The area of the Downs will also have been reduced slightly due to private development and quarrying in the intervening period. Thanks to the protection provided by the 1861 Act, the extent of the Downs has not been reduced significantly since that time.

While the common itself remained largely intact, the history of the Downs' physical space is dominated by two major and interrelated themes: urbanisation and encroachment. As landscape uses shifted from agricultural to recreational pursuits, rationalisation of boundaries occurred and some encroachments went unmanaged. Preservation of the Downs' boundary remains a contentious issue with local users keen to ensure that further encroachments do not negatively impact the green space.[12]

It is impossible to examine adaptations to Clifton and Durdham Downs without considering the development of the built environment surrounding the green space, not least the attempt to turn Hotwells into a spa resort. The spring had been used by wealthy invalids since at least the end of the sixteenth century and the Society of Merchant Venturers made efforts to purchase it from 1661.[13] During the late 1670s and 1680s, it gradually purchased the manor, and in 1696, it leased a section of it to two of its members with the aim of constructing a pump room so that they could tap the spring water from St Vincent's Rock below. The traveller Celia Fiennes described the water as being "As warm as new milk" and having "much of that sweetness".[14]

In 1710, the population of Clifton was said to be 450.[15] This increased dramatically with the building that accompanied the development of the spa. Residential housing, hotels, shops and offices increased in number in the area surrounding the green space, with developers capitalising on the desirability of this now fashionable suburb to wealthy Bristolians wanting to move out of the overcrowded city into new Georgian terraces. By the late 1720s, Hotwells boasted a playhouse and a Long Room with pleasure grounds, and building had begun around Dowry Square.[16] In 1746, a Sea Wall was constructed by John Wallis to protect visitors to the Spa from the cliff edge. The wall was capped with blocks of black slag to make it visible.[17] In 1750, the Merchant Venturers advertised 11 properties on Clifton Hill that were to be let or sold, reflecting the extent of urban growth in the area by this point.[18]

Despite the development, the Downs remained largely agricultural in appearance during this period and its popularity as a venue for recreation was reliant on its rural character. Depictions of the area from the early nineteenth century, including Francis Danby's *The Avon Gorge* and Thomas Leeson Rowbotham's *Panorama of Clifton* (1830), reflect this. Yet, the increase in visitors did spur a series

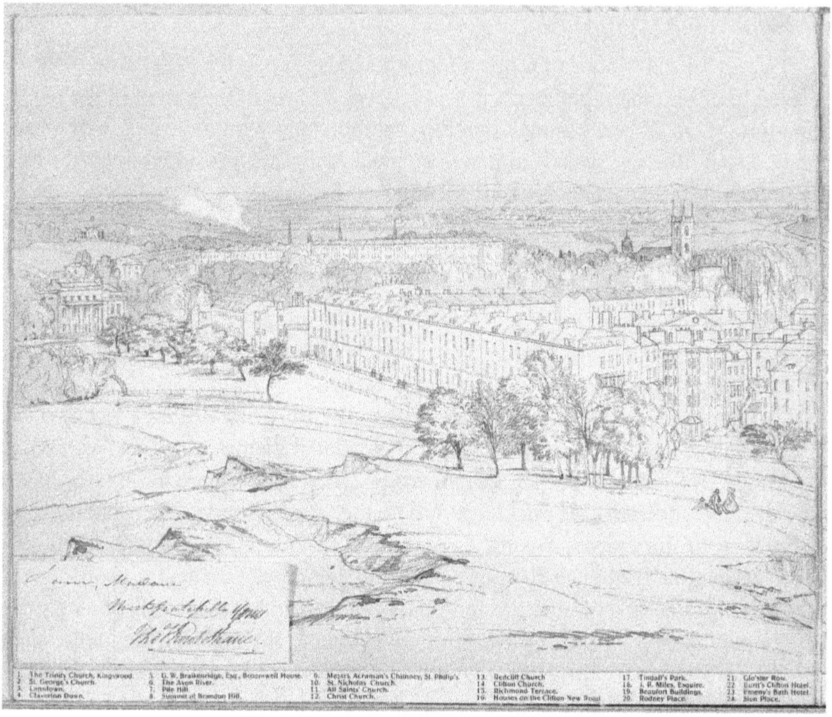

Figure 4.3 A section of Thomas Rowbotham's *Panorama of Clifton.*

of improvements such as the building of paths and roads. Rowbotham's sketches also show evidence of small, enclosed areas with railings which were presumably designed to protect recreational space adjacent to the new housing from grazing animals.

There are some indications that the Society of Merchant Venturers tried to prevent exploitation and encroachments occurring on Clifton Down through strategies like the appointment of a manager, Sir William Draper, in 1766. However, Draper permitted the levelling of sections of the hillfort, causing irreversible damage. In addition, there were competing interests between the community, local action groups (such as the resident-led Clifton Improvement Association, 1849–1855) and quarry owners, which resulted in various problematic decisions that impacted the surface of the Downs through extraction or inappropriate planting schemes.[19]

In the early months of 1859, the Society of Merchant Venturers was made aware of possible abuses of encroachment and enclosure occurring on Durdham Down.[20] Mr Samuel Worrall, whose ancestor of the same name had been Bristol's town clerk between 1787 and 1819, had erected a walled boundary at his property. The Society questioned Worrall's right to enclose a piece of land that had previously been open to the Downs, and although Worrall's right was upheld, the investigation of the case raised wider concerns about encroachment and led to the eventual protection of the Downs through the 1861 Act. Despite this, disputes rumbled on throughout the late nineteenth century over the willingness of the Downs Committee to submit to the demands of influential individuals such as Worrall.

Roads and walkways were another source of potential encroachment – and controversy. In the early eighteenth century, the installation of tollgates on Stoke Road and Westbury Road, which ran across Durdham Down, prompted a considerable backlash.[21] As the dominant use of the Downs shifted from grazing to recreation, new roads and walks were created to facilitate exploration on foot, on horse and by carriage. In 1822, Bridge Valley Road was laid out in the south-west corner of Clifton Down as a carriage drive to provide better access to the spa. Seven years later, the Zig-Zag Walk was created to improve access between Clifton and the Spa (which had originally been via a steep flight of steps). In 1862, controversy erupted when William Baker attempted to lay a new road across the middle of the Downs for the benefit of local residents.

Committee meetings between 1869 and 1875, chaired by Alderman Proctor, provide a useful case study on the continued debates that surrounded development of the Downs.[22] In 1869, a set of new proposals advanced the ideas of Proctor who understood that the abandonment of the tollgates in the 1840s had significantly increased traffic over the Downs. Proctor was in favour of a new scheme because he saw the benefit of encouraging people to visit the Downs "for the express purpose of loitering about and enjoying the beautiful prospects".[23] However, the Committee also recognised that without a scheme to control traffic and pedestrians, the green space would be subject to unauthorised roads and tracks that would cause harm. The adapted scheme eventually secured approval in 1875. It consisted of a two-mile circular road to be financed by subscription. Aesthetic concerns that the view across the Downs be maintained led to a decision that there would be no railings or planting along the road. Instead, new bye-laws were introduced to prevent people from driving off the road and onto the turf.[24]

In the twenty-first century, negotiations of access by car and foot still impact on the landscape of the Downs. Cars line the roads despite the attempts of the Downs Committee to protect the uninterrupted views of the green space, and new unofficial routes are created by the repeated footsteps of users who abandon existing footpaths to take the most direct path home.

Social and Political Context

Agriculture and Industry

For centuries, the commoners of the manors of Clifton and Henbury had the right to graze animals on the Downs. This practice was in decline by the mid-nineteenth century, and by 1925, grazing had ceased completely.[25] Today, the University of Bristol exercises commoners' rights to graze sheep on the Downs, which it does periodically. Since June 2011, goats have also grazed in the area of the Gully to control the growth of brambles and protect the rare plants of the gorge.

Quarrying and mining developed alongside grazing as a key land use. Early extraction tended to be small scale, but there is evidence of more sustained quarrying in the area of the Dumps. Earthworks dating back to the seventeenth and eighteenth centuries recall the extraction of lead, iron, manganese and calamine.[26] There is also evidence of an early modern limekiln at the foot of St Vincent's Rock that was in use by at least 1731.[27] Historic Environment data from Bristol City Council suggests that the presence of just one kiln at this location is indicative of a small subsidiary industry for the production of quicklime.[28] A local form of quartz was mined along the Avon Gorge and sold as "Bristol Diamonds" to tourists visiting the spa. By the nineteenth century, there were at least three large quarries on the Downs. One of these, situated between Stoke Road and the Centre Road, is visible in a watercolour by William Arnee Frank from 1865. Another, called Black Rock Quarry, was opened in 1868 to provide material for road building. Around the same time, work was undertaken to infill quarries that were no longer in use with a tramway constructed to transport spoil from building work at Bristol docks up onto the Downs.

A windmill was erected within Clifton Down hillfort in 1766–1767; this was a snuff grinding mill used to grind tobacco into snuff. The work was carried out by James Waters with the financial backing of the Society of Merchant Venturers. It was damaged in a gale and then burnt down in 1777.

Recreation

The use of the Downs for recreational purposes became increasingly common with the rise of the spa in the late eighteenth century. Samuel Grimm's sketch of Clifton camp from 1789 depicts horse-drawn carts on the Downs, and in 1819, Pierce Egan described how donkey-drawn curricles carried invalids onto the Downs for fresh air.[29]

There is evidence of horse racing on the Downs from at least the early eighteenth century.[30] In 1718, Sir Edward Longueville was killed after falling from his horse during a race, and a newspaper article from May 1749 described the extensive preparations prior to the races: the levelling of the ground; enlargement of the course; and construction of booths and scaffolds to accommodate racegoers. One account refers to races designed for spectators including some run under surprising conditions: "on Wednesday began the foot races, when 3 gs. were run for by two men, naked; and a Holland smock and one guinea by five women".[31] The races appear to have become an annual event from May 1828 and were depicted in a painting by Rolinda Sharples.[32] However, local residents complained, eventually securing the end of racing on the Downs in 1838.

Wrestling, cockfighting and track races also occurred on the Downs during the eighteenth century. A duel was even fought between Sir Henry Kippincott and R. B. Ward. Duels were illegal, but as the pistols were fired into the air, neither man was prosecuted. In 1824, another disagreement was fought out on the Downs, this time in a fistfight between a master baker and a butcher over a game of dominos. The baker, who was deemed to have the advantage, had his right arm tied behind his back. Even so he prevailed, with the butcher forfeiting after a close 15-round contest.[33]

More ordered and sedate games were also played. There is evidence of the Clifton Cricket Club playing matches on the Downs from as early as 1819 and a nine-hole golf course was laid out either side of Ladies Mile in 1889–1890.[34] A Downs Cricket League was established in 1902 and a Downs Association Football League followed soon after. By March 1930, it was reported that each Saturday there were 25 soccer, four rugby, three lacrosse and five hockey pitches prepared. Many of these activities continued until the outbreak of World War Two, but while football matches are still played on the Downs, neither cricket nor golf returned after 1945.

The playing of organised sports prompted calls for facilities. The first recorded request to the Downs Committee for permanent changing rooms came from Bristol Rugby Combination in 1903. The granting of permission was hindered both by the cost and the prohibition on erecting permanent structures on the Downs in the 1861 Act. To get around the problem, Bristol Water Works Company provided land adjacent to their reservoir, which had been purchased before the 1861 Act. This offer, coupled with a grant from the Parks and Open Spaces relief scheme, made it possible for changing rooms to be constructed. Following an application to Parliament in 1931, dressing rooms and washing facilities were brought into use for the 1932–1933 season.

Other buildings predating the 1861 Act facilitated different kinds of recreational activity. The road to the Ostrich Inn on Durdham Down is marked in the north-east corner of G. H. Hammersley's map from 1746.[35] The inn was said to be well known for its cockpit and bowling green.[36] In 1790, the publican erected lamps on the Downs which were lit at night to help customers find their way. In 1828, William West leased land on the site of the former windmill to build an observatory, an idea that had originally been proposed by Robert Pigott in 1789.[37] This initially held a telescope, which was soon replaced by a camera obscura. This device was used by the artist Thomas Leeson Rowbotham to produce his *Panorama of Clifton* of 1830.[38] The observatory proved popular with visitors, and West's lease was extended for a further 21 years in 1834 and he added a domed extension.[39]

Innovations such as the observatory reinforce the idea of the Downs as a recreational area for local upper- and middle-class residents rather than for Bristol's poorer inhabitants. One commentator noted: "It would take us an hour's walking, after the hard toil of the day is over, to get to these beautiful spots, and then another to get home". The arrival of the tram in 1875 opened up opportunities for a wider range of citizens to use the Downs for recreational purposes.

Fairs, shows and parades do seem to have attracted a wider audience to the Downs. The Bath and West of England Agricultural Show was held on the Downs on several occasions between 1863 and 1913. The comments of one American visitor in 1878 suggest that it attracted a socially diverse range of visitors. It was noted that farmers' wives and even female domestic servants wore dresses similar to their social superiors, with only their speech betraying their class.[40] The visitor also commented on the alcohol consumption: "men and women – of many classes, too – crowding about the numerous large booths where beer and spirits were sold".[41] The popularity of the show seems to have increased over the years, with the area it covered being extended from 25 acres in 1763 to 70 acres by 1878. In 1913, King George V was in attendance and Bristol Council arranged a procession from the city to the Downs with the King travelling in an open-top carriage.

King George was himself the focus of the Jubilee Fair in 1935, which included a fairground on Clifton Down. The carnival that took place on Durdham Down two years later was so popular that it had to be extended by a week and takings reached over £5,000.[42] The event opened with an "Exotic Oriental Scene", involving a display of 200 multicoloured Chinese lanterns performed by members of the Church Lads Brigade. The carnival also hosted naval and military displays and 20,000 tickets were given to local children for rides on "roundabout machines".[43]

Some of the fairs were about more than entertainment. In 1910, Sir George White organised flying displays using box kites to advertise his Colonial Aeroplane Company. In 1936, the NSPCC used the Downs for a promotion event at which gymnastic displays were accompanied by information about the organisation and its mission.[44] During World War Two, the fairs were replaced by the "Holidays-at-Home" scheme, which included a circus. In 1942, an eight-year-old boy was taken to hospital with serious injuries, having been mauled by a lioness.[45] The fairs returned in the aftermath of the war with a festival on the Downs in 1946 – involving sports, choirs and folk dancing – organised by the Bristol Co-operative Society,[46] annual horticultural shows from 1947, Jubilee celebrations in 1977 and 2002 and Bristol 600 in 1973 commemorating the anniversary of Bristol's incorporation.

The Downs is still widely used for recreation, and this was reflected in the oral history interviews for this case study. During the COVID-19 pandemic, the Downs saw a huge increase in the numbers of people using the space. There has long been a heavy emphasis on learning and education, with the education officer co-ordinating a number of classes, workshops and walks with experts on various aspects of the Downs. These include butterfly surveys, litter-picking groups, volunteering, art classes, tree walks, moth walks, children's groups and creative writing. During a lengthy interview, a member of the education team explained the contemporary benefits to this kind of engagement:

> … the education community engagement programme is all about raising awareness about the importance of the Avon Gorge, Downs and Leigh Woods, and also creating opportunities for people to come and enjoy, discover, learn about the special wildlife and landscapes … we totally believe in the ethos that the Downs is here for everybody in Bristol. So, we do want as many people as possible to experience it and build a connection with it, really. Because, you know, it's always well said that wildlife and landscapes don't get protected unless people know what's there, they get to love it and then they value it and treasure it. If it's at any stage under threat, not that the Downs are, but, you know, kind of like it enables people to actually have a voice because they're well informed as to why it's special, both in the wildlife but also from their personal perspective, you know, their memories of visits and their family events that have happened on the Downs, you know, that all gets tied into why they love a place. [47]

The education engagement project began in 2001. The programme looks at the contemporary Downs, as well as the history and "the underlying geology" of the land and how it has traditionally been managed.[48] In order to increase the scope of the projects, Downs management looked beyond the immediate Clifton area and has worked with groups throughout Bristol, including refugee groups, mental and physical health charities and schools from different parts of the city. One oral history interviewee explained that these initiatives were designed to welcome people to Bristol, but also to "deepen their connection with nature" and the community.[49] In the focus groups held after the events, participants emphasised that "being in nature made them feel happy and relaxed" and "that the beauty of nature kind of supported their optimism and sense of wellbeing". One participant made this poignant comment: "I loved watching the birds on a tree. I learnt that life is not only war; life is very beautiful".[50]

For one interviewee, the history of the Downs influenced their perspective, and as a writer, photographer and local resident, they argued that Bristol's transport system was "dated", which prevented those on low income accessing various parts of the city, and the Downs themselves had a reputation for exclusion:

> … I completely, personally, believe that when the Merchant Venturers bought the Downs, it was 100,000% to stop it being urban sprawled over. Right? So, I don't think it's a commons land … I know that it was commons. But I don't see it as commons now. And I don't see it as a… I don't see it as like Roundhay Park in Leeds … or Hyde Park in London or… or tiny parks nearby that were given by like the Smyth family or Ashton Court which was given to the people of Bristol in perpetuity. I see it as just like this symbol of wealthy people don't want the proles living near them and want a nice place to go and walk.[51]

Interviewee 2 believed that "we want to keep the proles out" was "written all over the landscape" because of this inaccessibility. Importantly, s/he felt this inaccessibility was ingrained in the history of the Downs, which had been designed to exclude the less wealthy. Interviewee 1 also recognised these barriers and the dearth of transport options for low-income Bristolians. She also acknowledged a "mental challenge" around the Downs because of their location near Clifton and Redland, two wealthy areas of Bristol. S/he explained that this often led to potential visitors worrying about whether they were allowed to use them: "I think it's quite often seen by some people on the other side of the city as being, 'oh, that's not for us. That's where posh people go'".[52] She also acknowledged that the education initiatives worked well with some families because it addressed how to enjoy a green space. The workshops revealed that some parents were unsure of what to do with their children there or even fundamentals like what to wear.[53] Interviewee 2 also highlighted this by asking, "how are we taught how to use green spaces … when it's something that's like as weird as a big, open field, you know, a big piece of nature in the middle of a city?"[54] These interviews raised excellent questions about who is not using urban commons, and why, as well as how historical perceptions of urban green spaces can influence contemporary use.

There were also frustrations about the balance between maintaining the land and encouraging users. One local resident explained that the environmental initiatives had to work in tandem with the recreational activities that take place on the Downs:

> Wildflower meadow[s], orchids and many other things. It's very important. And if the runners, and there are lots of runners–, don't get me wrong … running is such a good thing … but there are more and more runners because of Covid. And if they all run through the wildflower meadow, sorry, that's trashed.[55]

This was echoed by volunteers who ran and participated in litter picking on the Downs. Interviewee 4, another local volunteer, explained that the Down's management kept it "looking absolutely remarkable", especially considering the number of people that used it during the summer of 2021, yet it looked "great when you arrive, and it looks appalling when you leave".[56] S/he argued that this damage was largely "down to young people" and that the best way to combat this was to educate children about green space in schools: "If they were taught in schools that, you know, the damage that litter can do, I think more of them would be more thoughtful about clearing up after them".[57] By contrast, Interviewee 5 considered whether the litter came from another source:

> I don't think it is [young people]. I think it's the bankers. It's people with lots of money to spend on prosecco, have a good time, all this bought food, plastic wrap and so on. It's a difficult one, isn't it? I don't think it's the kids, myself … Because some people leave litter not because they want to sort of give two fingers to society. They just don't think. And it's a very difficult one. I think primary-school kids particularly, and in the younger secondary kids, are much more aware of environmental issues than the generation before them. It's only my feeling.[58]

It was clear that the local residents and Downs volunteers were concerned about marrying the increase in foot traffic with the need to preserve the endemic species and beauty of the space. Regardless of the cause, it was clear that the recreational programmes were not only popular but also designed to educate the public on the environmental importance of the Downs.

Military and Political Uses

Clifton and Durdham Down have long been used for military and political purposes. In 1643, they served as the assembly point for the royalist army that captured Bristol. Two years later, they were used by the Parliamentarian forces, intent on recapturing the city and preventing the escape of the royalist commander Prince Rupert. As Joshua Sprigg explained: "Commissary General *Iretons*, Colonel *Butlers* and Colonel *Fleetwoods* Regiments of Horse, were appointed to be in a moving body upon *Durdham-Down*; that place being the most open way, and most likely for the Prince to escape by".[59] In 1796, the Hampshire fencible cavalry assembled on Durdham Down for the consecration of the colours; in the mid-nineteenth century, the Duke of Beaufort's Regiment of Yeomanry used the area near the Sea Wall for their annual training exercises; and in 1862, a military review was assembled on the Downs involving 8,000 volunteers and 100,000 spectators. The area continued to be used for inspections during wartime. In 1941, the Post Office Home Guard paraded there, and in 1943, 1,000 firemen and women marched before Commander Kirkup and Bristol's Deputy Lord Mayor.

In the times of war, the open expanse of the Downs presented both risks and opportunities. Measures were introduced to deter the enemy from landing on the Downs, including barrage balloons, anti-aircraft guns and 41 stone cairns. The space was also used more positively with the erection of telegraph posts,[60] the establishment of a tank repair depot close to the Sea Wall and use of part of the Downs as a military vehicle park.

The threat of war also brought peace protestors to the Downs. In 1933 and 1937, anti-war demonstrations were organised on the Downs. In 1933, speakers condemned the expenditure on arms and the attempt to create a "war spirit" through military displays. A further anti-Fascist, pro-peace meeting was held in 1938.

Political meetings on the Downs were not new. The Chartists had gathered on Clifton Down in the mid-nineteenth century; a Trade's Congress rally was held there in 1915 to highlight the dangers of labour on the railways; and in 1921, there was a mass meeting to protest at high unemployment levels.

Crime and Death

Given the range of activities taking place on the Downs, it is not surprising that crime and death occasionally occurred. The most common crime was robbery. During the eighteenth century, several men were executed for robberies that had taken place on the Downs.[61] The problem continued into the nineteenth century. In 1811, a gentleman was robbed of his money and watch at gunpoint while he sat reading near Observatory Hill. By 1819, one newspaper reported that the streets surrounding Clifton Down were "so infested at night with desperadoes" that "few gentlemen" thought "it safe to walk about alone".[62]

Though less common, murders also took place on the Downs. Perhaps the most notorious was the case of nine-year-old Matilda/Melinda Payne in 1855. She had been sent across the Downs by her father to fetch a jar of beer, but never returned. The following day, her remains were discovered and a bloody knife was recovered from a hiding place in the Zig-Zag Walk. A laundress described having encountered a man acting wildly in the early hours of the morning. He had blood on his coat and was muttering about "Hell's Channel". Around this time, other women and children reported being frightened by a man on the Downs. Deaths also occurred as a result of occupational hazards faced by those working on the Downs, such as the miners who were stifled by smoke in 1574 while digging for tin and lead. Several cases of suicide were reported, including that of a man who slit his own throat on Clifton Down in 1853 and another who was found on Durdham Down in 1930 having consumed a poisonous substance. In 1934, May Muir Kerr deliberately drank disinfectant "in full view of holidaymakers on Durdham Down".[63] Kerr was a patient at Dorset House, a residential clinic for women with mental health disorders. It was located near the Downs precisely so that patients could benefit from time in the open air.

Governance and Property Rights

Contemporary Clifton and Durdham Downs manifest closer affinity with the manorial origins of commons than the other case studies presented in this work. Clifton and Durdham were separately registered as common land in the Commons Register for the city of Bristol under the Commons Registration Act 1965.

The modern governance of the Downs is based on the distribution of rights and responsibilities laid down in the Clifton and Durdham Downs (Bristol) Act 1861. This empowered the Corporation to purchase the freehold of Durdham Down for £15,000. The 1861 Act established governance arrangements based upon a joint management principle whereby both Clifton and Durdham Downs were henceforth to be managed by a Downs Committee comprising equal numbers of representatives of the Council and the Merchant Venturers. The overarching responsibility of the Committee is to ensure that "the downs shall for ever hereafter remain open and unenclosed, and as a place for the public resort and recreation of the citizens and inhabitants of Bristol" (subject only to commonable rights over the Downs).[64]

The residual influence of the Downs' origins in the manorial system can be seen in the continuing existence of common grazing rights. The Downs committee must recognise the property rights of commons graziers who have a right to depasture sheep on Durdham Down; there was also a "wrangle right" (an undocumented right to pasture sheep) on Clifton Down prior to the introduction of commons registration by the Commons Registration Act 1965. There is evidence that as late as 1917, local graziers claimed to have the right to graze up to 1,875 sheep on Durdham Down – although whether this number was ever actually turned out on to the Down is open to question.[65] In the case of Clifton Downs, the commons register notes 230 acres of land owned by the Merchant Venturers registered as common land.[66] The registered common excludes the site of the former Clifton Turnpike House and the Clifton Observatory. Although Clifton Down is registered as common land, however, no common grazing rights are registered over this part of the Downs.

In the case of Durdham Down, on the other hand, the commons register records 19 separate entries of grazing rights for sheep held by a variety of individuals and bodies, including Bristol University and the Bristol City Council.[67] These are all "appurtenant" rights, in that they are attached to and benefit parcels of land in the vicinity of the common.[68] The largest right registered was to pasture 655 sheep on the common, held by Clifton Theological College and attached to (benefitting) Stoke Mansion House, Stoke Bishop. In all, common rights to graze 1,885 sheep on Durdham Down were registered under the 1965 Act, and these rights persist to the modern day. In practice, however, they are rarely exercised. Following an outbreak of sheep scab in 1924 and a prohibition on grazing on the common issued by the Ministry of Agriculture in 1925, the commoners annual meeting agreed in May 1926 not to depasture sheep on Durdham Down.[69] Grazing has not been practised as a commercial enterprise since then. However, sheep are turned out onto the common periodically in order to maintain the common rights, and in some cases Bristol University permits its flock to be used in this manner. The commoners meet annually, and at least one sheep is tethered on the Downs for a day every 10 years.[70] Bye-laws made under the 1861 Act require the permission of the Downs Committee to be obtained before animals can be grazed on the Downs.[71]

Shaping Contemporary Clifton Downs: Governance of the Downs

Contemporary governance is overseen by the Downs Committee under the rules establishing a joint management principle model set out in the 1861 Act.[72] These continue in force and seek to implement a balanced approach to the management, preservation and development of the Downs under the aegis of a single management body whose actions are shaped by the interests of the Council, the Merchant Venturers, commoners and the wider public who use the Downs for a variety of recreational purposes. The 1861 Act established a 14-member Downs Committee comprising the mayor and six members of the council (elected by the council) and the Master and six members of the Society (elected by the Society). The joint committee serves until 9 November in the succeeding year, at which point a new committee is constituted. Each committee member is elected for one year, and the 1861 Act stipulates that two of the members nominated by the society and two of those nominated by the Council are not capable of re-election. The mayor acts as Chairperson, and in his/her absence the Master of the Merchant Venturers takes the Chair. This carefully balanced approach is representative of the bodies' respective property interests in the Downs. But in a modern context, it is arguably problematic, in that it hinders continuity of decision-making and planning by the committee – there is some continuity of membership, but the requirement for four members to stand down each year, together with the arrangements for annual elections of both the mayor and the Master of the Society, means that in practice the committee has a shifting membership. This leads, it has been argued, to a lack of strategic direction in management and forward planning and is insufficiently flexible.[73] The arrangements have been described as "a wise, albeit conservative, mechanism for decision making".[74]

The Council has the power to make bye-laws for the purpose of "regulating, improving, and preserving Durdham Down and … Clifton Down for enforcing such orders as the Joint Committee may make".[75] Bye-laws and any amendment or revision require the approval of the Society. The bye-laws introduced following the passing of the 1861 Act reflect the significance and importance of the Downs as an area of recreational public access. Commercial activities that could interfere with orderly recreational use were prohibited from an early date. For example, a bye-law was introduced in 1892 prohibiting carpet beating on the Downs.[76] The current bye-laws were introduced in 1998 and amended in 2003. These prohibit the use of motorcycles, motor vehicles, bicycles and "any other propelled vehicle" on the Downs other than in areas where there is a right of way for that class of vehicle.[77] They also prohibit the erection of any building, fence or other structure on the Downs without the Downs committee's permission or the removal of any structure.[78] There are also provisions in the bye-laws to protect flower beds, shrubs and areas of grass renovation, for example, by prohibiting the parking of vehicles, bicycles and more on them so as to damage them.[79] These restrictions are fairly closely aligned with similar restrictions seen elsewhere for the management of "people's parks" established in the Victorian period, and reflect the importance attributed to the Downs as a "green" space close to the urban centre of Bristol to be maintained as an area open for public recreation. It is perhaps noteworthy that there was no formal proposal in the period leading up to the 1861 Act to establish the Downs as a "People's Park" as such.

In the contemporary context, where public spending is severely constrained, the problem of financing the ongoing management of the Downs has acquired some prominence. The costs of the Downs committee and management are covered by the Council. The 1861 Act permitted the Council to charge a precept up to £300 a year to cover the acquisition and management of the Downs.[80]

Shaping Contemporary Clifton Downs: Environmental Governance and Management

Following a public consultation exercise, a Management Plan for Clifton and Durdham Downs covering the period 2012–2017 was established in January 2012, with the objective to assist the Downs Committee in long-term planning for environmental improvement of the Downs.[81] This builds on – and links with – the Council's *Parks and Green Space Strategy* (2009), which sets out a 20-year investment programme to establish and maintain good quality, attractive green space that is accessible and meets the needs of Bristol's citizens.[82] One of the key challenges addressed by the Downs Management Plan is the problem of encroaching scrub that has become more prevalent following the cessation of sheep grazing in 1926. The Avon Gorge and Downs Wildlife Project was launched in 1999 and has, for example, undertaken a limestone grassland restoration project to link existing areas of natural grassland and create a larger and more viable area, and introduced a herd of feral goats in 2011 to help restore wildflower-rich grassland and control scrub encroachment.

The Downs Management Plan included a five-year work plan to manage and improve the physical and cultural environment of the Downs.[83] The Downs contain large areas of unimproved and semi-improved calcareous grassland. The unimproved grassland is managed as a hay and wildflower meadow cut annually to promote the seeding of wildflowers.[84] Further to the management plan, a tree planting plan was prepared to allow for the replacement of trees in avenue plantings with existing or new species.[85] And the management plan set out several initiatives to improve biodiversity and wildlife on the Downs; these include carrying out a survey to investigate the options for increasing the areas of wildflower meadows in selected locations and undertaking a grazing feasibility study.[86] Part of the Downs extends into the Avon Gorge, and this is an area of very high nature conservation value. The Gorge has its own management plan, which links with and complements the Downs Management Plan itself. Its botanic importance is considerable – it hosts no fewer than 27 nationally rare plant species linked with limestone grassland communities.[87] The Avon Gorge has been recognised since 1952 by its designation as a protected area under national environmental legislation.[88] It was re-notified as a Site of Special Scientific Interest under the Wildlife and Countryside Act 1981 in 1988, its primary protected features being its rare plant assemblages and its geological interest. The Sites of Special Scientific Interest (SSSI) rises about 100 metres above the River Avon to Observatory Hill and to Stokeleigh Camp in the west and includes part of Leigh Woods.[89] The site notification sets out Natural England's views about site management, and these include guidance on protecting the geological features of the screes and outcrops on the Gorge sides by active management (including light levels of livestock grazing to suppress scrub encroachment) and on maintaining a species-rich sward on the grassland areas of the Downs – again with grazing.[90] The SSSI management guidance informs the approach adopted in the Avon Gorge and Downs Wildlife Project and the Downs Management Plan. The Avon Gorge Woodlands are also designated as a Special Area of Conservation[91] and Leigh Woods is a National Nature Reserve.

The Downs today exist as both a recreational space and a space concerned with preserving the habitats of endemic wildlife and other environmental factors. As a group, the Friends of Downs and Avon Gorge (FoDAG) aimed to contribute as much as it could to "restore some of the balance to the natural world, to the advantage of the natural world rather than to just satisfying people's perceptions of what a green space should be like".[92] As with the educational programmes, the FoDAG recognised that there was public interest in the Downs. This meant that, potentially, anything other than planting trees or maintaining the status quo was a cause for concern, but with expertise and education, this does not have to be the case. One local resident, and member of the FoDAG group, gave the example of the dead hedging barriers being constructed on the Downs:

> People walking past say, 'What are you doing?' and you explain about nudging the runners out of the wildflower meadow and they say, 'Oh, that's a good idea. Well done. Thank you'. And that sort of continues the education if you like.[93]

By maintaining public interest and visibility with regards to management as well as robust volunteering programmes, the Downs management ensures that members of the public are enlightened as to the purpose of these alterations. One interviewee, a volunteer and expert on local wildlife, explained how members of the public often had misconceptions about the best way to conserve the land. He argued that "the environmental and climate change issues at the moment tend to get pushed into these very simple tropes" which lead to the prioritisation of tree planting:

"More trees, more trees, more trees, that's going to save us when the world turns to fire". One interviewee explained that reforesting was a crucial part of a global conservation strategy, but that this needed to happen on the green deserts that already exist and not on the extensive patchwork of habitats on the Downs, which were themselves an important ecosystem.[94]

He also underlined the importance of this strategy. He argued that one of the benefits of his retirement was the ability to volunteer on the Downs, because there is "just so much to interest and do in terms of protecting the wildlife, conserving it and developing it in ways that make it more accessible to the general public".[95] He explained that this accessibility was important because it allowed for more conscious decisions to be made about the landscape:

> In a way it is, yes. I mean one of the first things that was done up there, this was before I got involved, this was in the 1990s, was they started after a lot of hard work persuading people to leave areas of the Downs grassland uncut and just harvested it in the autumn which meant that natural meadow was re-established. Because the use of mowers to cut the grass was, as with so many parks, about making it look tidy. And nature isn't tidy.[96]

He argued that the Downs was an example of how to "get things working and restore some of the balance to the natural world … rather than to just satisfying people's perceptions of what a green space should be like".[97] This was an important reflection. As with Mousehold Heath and in contrast to the perceived monoculture of Town Moor, the Downs management places importance on a variety of habitats and ecosystems, which can exist in tandem with recreation. This means that the prescription "to remain open and unenclosed" is adhered to, but that the landscape can also flourish.

Present Voices Future Directions

The well-established education programme that supports the engagement of Bristolians with the Downs, its history and environment, acts as a productive model for future public engagement in urban commons more widely. Equally, the education team at the Avon Gorge and Downs Wildlife Project proved decisive in setting up participatory research activities for the Wastes and Strays research project, facilitating connections to local schools and community groups, overseeing participant recruitment and managing workshop practicalities. Participatory research for the Bristol case study comprised of creative writing workshops with two school groups and the public to explore historical research and contemporary experiences of the Downs,[98] along with self-directed creative explorations and walking conversations carried out with community participants. The latter were restricted by the COVID-19 pandemic lockdowns, which reduced public contributions to the future strand of this research project. In support of this research, the Wildlife Project's education team acted as gatekeepers to local communities in a way that illustrates the potential for other organisations working in the fields of the environment, arts and health to gain crucial support in the delivery of community-centred projects, which was not evident in the other case studies.

The interrelation between connectedness to an urban common environment and the future action of citizens to protect the site emphasises the value of nature experiences promoted by the Avon Gorge and Downs Wildlife Project. Recalling the comments made by the education manager, which were articulated earlier in this chapter, the knowledge of and bond to the Downs "enables people to actually have a voice" because they are both informed and personally invested.[99] Connectedness is not simply a matter of going out for a walk or a run but emerges in engagement opportunities that foster both cognitive and affective relations with the downlands, or rather, ecological knowledge and emotional connections to nature.[100]

An individual-scale example was tendered by a participant in an account of his involvement as a volunteer with the Friends of the Downs and Avon Gorge. "Up until 2010, I knew nothing about butterflies, I'd never really given them a thought", Tim explained.[101] Yet, participation in a one-off butterfly transit walk led to years of butterfly monitoring, and through this engagement he had accrued the knowledge to identify species:

> It's always disputed how many species we see but I reckon there's about 30 different species which out of, there's only 59 British butterflies so you know we get over half of them in this area.[102]

Furthermore, the affective relations engendered in this getting to know the ecology of butterflies seemed to inspire Tim's response to the question of what aspects of the Downs were of highest priority to preserve into the future:

> I suppose it would have to be the butterfly areas really, just making sure we are still getting the regular butterflies. So, you know, protecting the wildflower meadows.[103]

Tim's comments serve to illustrate how a combination of ecological knowledge, as demonstrated in species identification, and affective relations with nature, expressed in the "satisfying" feelings of nurturing those species, can influence people's sense of care for an environment. This is not to suggest that changing behaviours is simply a matter of straightforward cause and effect strategies; however, there is growing empirical evidence of the interrelation between nature connectedness and pro-environmental attitudes and behaviours.[104] If such evidence is applied to the urban common, the proactive development of citizens' learning and engagement may help to mitigate the disconnect from nature typical of urban dwelling and ultimately support conservation into the future of the commons and more broadly.

In a public consultation on the newly proposed principles and workplan of the Downs Committee carried out in 2022, respondents placed a slightly higher value on the Downs as "a space for recreation" than "a space for nature conservation" by a margin of 7%.[105] The priority placed on educational activities as the type of event respondents most wanted to see in increased provision is interesting,[106] then, because learning experiences can bridge the two priorities of recreation and conservation. Moreover, the learning programme supports the committee towards fulfilling its statutory commitment to serve all Bristolians. Since the programme began, the education manager explained, "nearly 130,000 people have come on one of our walks or talks or been on a school visit or playscheme visit or they have taken part in one of our community-engagement programmes".[107] Furthermore, drawing up an audience development plan in 2018 enabled the education team to identify audience gaps and target potential partnerships, which resulted in the expansion of provision to communities and schools across Bristol, including "from more disadvantaged areas of the city".[108] It is noteworthy then that a majority of respondents across all categories – "members of the public who use the Downs for recreation/ exercise", "residents living next to the Down" and others, including "non-users" – all identified there were "too few" educational activities.[109] In the survey, 57% of people felt there were "too few" educational activities, as exampled by guided walks and education projects, which can be compared to 17% who identified "too few" sport events and 6% who wanted to experience more circus.[110] The public aspiration to experience the Downs as a site for learning is already informing the Downs Committee's plans. A recent proposal to restore areas of hay meadow, including the former zoo car park beside the Ladies Mile Road, will involve school children participating in conservation tasks, distributing green hay across the meadow, planting seeds and plug plants.[111]

In the recent drafting of guiding principles, the Downs Committee has placed greater emphasis on the downlands' role in nurturing citizens' well-being. The purpose of the Downs is prescribed in the 1861 Downs Act as a "place of public resort and recreation";[112] however, the principles advance a new vision:

> An open downland that is protected forever for the wellbeing and enjoyment of the people of Bristol.[113]

Where the commitment to maintain an unenclosed downland is restated, its value for well-being reflects contemporary thinking on the interrelation between access to green space and city dwellers' physical and mental health. Developing initiatives to support well-being is another aspect of the programme delivered by the Avon Gorge and Downs Wildlife Project, as identified by its education manager:

> We've done a lot of work with the five ways to wellbeing walks for the Bristol inner-city health improvement teams.[114]

Through the partnership with the Health Improvement Team, volunteer walk leaders have been trained and regularly facilitate health walks, offering support and encouraging people to be more active. Tim is one such walk leader, although his facilitation of health walks on the Downs has spanned both his time as a mental health worker employed by the NHS and now as a volunteer member of the Friends of Downs and Avon Gorge (FoDAG). At present, Tim has around 30 people who regularly come out on his Monday walks, which is an increase "by a good fifty percent" since before the COVID pandemic.[115] Elsewhere in this research, a correlation has been drawn between the negative impact of the pandemic on well-being and the increased use of urban green spaces. However, Tim attributed the rise in attendees to the closure of walking groups that did not resume after restrictions were lifted. People who participate in the walks are referred by a health professional, either their GP, practice nurse or a social prescriber. In contrast to other public walks on the Downs, health walks specifically aim to address people's health needs and promote well-being, an intention that is also apparent in the application of the Five Ways to Well-Being.

The Five Ways to Well-Being is a set of evidence-based, daily practices to improve individual well-being developed in 2008 by the New Economics Foundation.[116] Framed as messages to prompt behaviour change, the approach suggests well-being can be accrued to people who spend more time socialising, being physically active, paying attention to the present moment, learning new things and giving to others.[117] Since the publication of the framework, national and local organisations have been promoting the practices to nurture positive mental health and well-being, alongside care and treatment.[118] The partnership delivering health walks on the Downs as an intervention to support people's mental ill health is a case in point. The Five Ways to Well-Being, then, provide a useful lens to reflect on the Downs' health walks, and more specifically, Tim's experience as a community walk leader.

The practice of walking in the Downs clearly prompts engagement in physical activity, but asked whether this was the reason people attended, Tim commented:

> I think most people come on the walk, it's the social thing, a chance to meet people.[119]

The opportunity for social interaction then seemed significant to people's engagement. Interestingly, group sociability also acted as a conduit for learning. The health walks "aren't guided", in the sense of a wildlife identification or heritage walk where an "expert" explicitly informs participants, but Tim observed, "people are going on them to build up a knowledge base of the birds, trees, flowers".[120] In this case, knowledge seemed to be accrued through informal sharing between participants and with a well-informed leader, although Tim confessed: "I'm not good on wildflowers".[121] Where noticing and giving were not referenced per se, the forging of social connections and discovery through social interaction suggest a relational curiosity rooted perhaps in paying attention and a sense of care.

In contrast, a participant in the research project's creative co-inquiries questioned the capacity of walking for health to necessarily foster greater mindfulness and belonging. Evaluating her participation in the self-directed creative explorations, Natalie Smith reflected:

> Something like this could be included in the 'social prescribing' model run by GP practices. It would have been far better for me to have done something like this, than (the walk around a cemetery that our little group did on my first GP/social prescriber session!).[122]

Natalie Smith had enjoyed the different modes of engagement with the natural environment offered in the self-directed walking explorations of the Downs: drawing, writing, observing, photographing, storytelling, reflecting:

> I have visited the Downs for many years so it's great to have been able to look at it through new eyes.[123]

Her reflection illustrated how creative practices had enabled fresh ways of looking at and encountering the downlands while out walking, an example that seems to align more fruitfully with a sense of paying attention to the present moment. Going into the future, both approaches of walking for health and arts for health are ripe for development in engagement programmes, such as those delivered by the Avon Gorge and Downs Wildlife Project, which aim to foster encounters with urban commons' environments to support mental health and well-being.

Figure 4.4 Natalie Smith, *Surface Collage.* In-Common Sites co-inquiry No. 4, the Downs, Bristol, 2021.

Image: Natalie Smith.

Returning to the Downs Committee survey, the findings clearly demonstrated public ambition to extend learning opportunities; however, direct comparisons with proposed alternatives are complicated by perceptions of adequate provision in other event areas and concerns over the increase in large commercial events. Walk leader Tim identified one such event that seemed to provoke division:

> There's always controversy about um; they do a big concert, pop concert in September time, which I would have no interest in, but it does bring in quite a lot of finance for the Downs. And it's only one weekend. So, I think it's fair. And they always do a very good clear up after it. You know, they leave it better than when they found it. So, I've got no issue with that.[124]

Where Tim acknowledged tensions over the programming of this two-day ticketed event, he also recognised the imperative to raise funds for the maintenance of the Downs. The issue of how to finance the maintenance costs of urban greens is of growing concern. Where the budgets of local authority landowners have seen real term decreases over the last decade, the provision of parks and green spaces is a non-statutory function, which leaves them open to cost-efficiency measures. Another observation from the Avon Gorge and Downs Wildlife Project education manager, whose salary is part funded by the Downs Committee, conveys a sense of the difficult decisions faced by Bristol City Council and the committee over funding the downlands:

> We had a very long, stable period of funding and then in 2016 the city council cut all sorts of funding. There was that, what did they call it where we went into like austerity, and it all just got cut. And so, for a while there was a massive panic because the money from Downs Committee got completely slashed. But that was also at the same time that there were discussions going on about green spaces in Bristol, because there's no statutory funding for green spaces, it's not like other services like, you know, housing or schools or roads. And so, it's kind of discretionary as to what local authorities spend on green spaces. And so, there was these discussions about making all green spaces self-sufficient. So, at that time we were having some big events on the Downs, but now we have more to bring in income to enable the functioning and running of the Downs.[125]

This reflection emphasised how the pressures of austerity on the city council's budgets placed constraints on funding with potential detrimental effect on the maintenance and investment in urban green spaces. The escalation in commercial events on the Downs then reflected Bristol City Council's plan for its parks to become self-sufficient in order to protect green space and ensure "free access" into the future.[126] Proposals set out by the council in a "future funding model" in 2018, included community involvement in the maintenance of green spaces and income generation from cafes, pay and display parking, fee-paying activities and events.[127] The enactment of the model in the Downs, however, is complicated by the 1861 Downs Act which binds the council to meet the costs of maintenance[128] and places legal constraints on certain revenue raising activity. Public users and community groups have therefore questioned the legality of the Downs Committee to execute events and activities, which seem to contravene the Downs' status as an open green space for the public at large to enjoy in general recreational use. But where ticketed, enclosed events are by nature exclusive, festivals such as the Future concert or Bristol Pride[129] and theme park, Funderworld do draw in audiences from across the city.

As the Act serves to protect the Downs from development, so too it enables local people to exercise civic stewardship by employing legal means to ensure the Downs Committee operates within its statutory remit. A prime example was evident in the extended controversy over the granting of temporary licenses to Bristol Zoo for visitor car parking on the Downs. Since 1996, local users have objected to seven planning applications and set up the group *Downs for People* in 2013 to coordinate campaigns against the zoo parking on the Downs.[130] In August 2020, the group mounted a High Court challenge to the Downs Committee's grant of a 20-year licence for a temporary car park, and in May 2021, an out-of-court settlement was agreed in the group's favour.[131] As a result, the Downs Committee and Bristol City Council "have given a legally binding undertaking that they will never again set aside land on the Downs for parking activities taking place elsewhere",[132] and, as outlined above, the meadow will undergo rewilding. The cost to Bristol City Council of the legal challenge, estimated by *Downs for People* at £250,000, though this figure is not verified, led the group to publicly call for a review of the committee, claiming it was "not fit for purpose".[133] Ultimately, the perceived pressure to make the Downs self-financing precipitated a costly expense on council taxpayers. Moreover, the public users' sense of stewardship for the downlands galvanised direct action, which consequently compelled the Downs Committee to engage in a wide-ranging review of the Downs and its governance.

It was, in part, these events, alongside the work of concerned local councillors, which led the Downs Committee to undertake a listening engagement event with stakeholder groups, the drafting of principles[134] and a work plan[135] and a public consultation on those plans.[136] Issues of concern raised in the consultation coalesced around the transparency and openness of decision-making in the Downs Committee; the preservation of the Downs as open green space protected from development; the balance between large fee-paying events, small and culturally diverse events and the purpose of conservation; and the financial obligations of Bristol City Council.

The resultant Principles of the Downs Committee have reaffirmed the council's responsibility for funding maintenance; however, this is coupled with an ambition to "save public money by raising revenue".[137] The issue of funding then remains prescient. In public consultations, stakeholders were asked for ideas on alternative funding sources. Some suggestions, for example, the development of a sponsorship programme, have been taken forward into the committee's workplan.[138] The proposal to develop the public toilets at the Sea Wall into a café, however, despite expressions of public support,[139] has proved unviable, following consultations with national government over the limitations of the Downs Act.[140] The proposition of funding the Downs through a community precept, as outlined by *Downs for People*, is of interest, for it has potential application to urban commons more broadly:

> In most of the country outside some large urban areas, including Bristol, there are parish/town/community/neighbourhood councils with unlimited precepting powers on residents in their areas, as long as those residents agree. Most put money into green spaces.[141]

The recommendation, then, would be to set up some sort of council with the power to levy a tax on residents to fund services within the local community, which in this case would be the maintenance of the Downs:

> A council could be established just to fund the Downs (at least in part) if neighbouring areas supported the idea. This would have the advantage of costs falling on those who benefit most and avoid any impact on other council services.[142]

The merits of the proposal seem to intervene in the long-standing conflict over the relationship between cost and benefit and how the Downs functions as a local or citywide asset. To be enacted, however, the motion requires the agreement of local residents to the setting up of a community council and the charging of a precept. The recommendation, therefore, encouraged Bristol City Council to test the viability of the proposal as part of a Community Governance Review.

Where the notion of a precept presents a longer-term, probably controversial, solution, current plans retained the focus on events as a means of income generation. In the public survey, people were asked to indicate their preferred method of funding for the Downs. Overall, a small 54% majority of respondents preferred the Downs to be self-funded through events and activities rather than funded by Bristol City Council.[143] However, this figure rose to 69% of the public who do not use the Downs and fell to 45% of residents living next to the Downs.[144] On the one hand, citizens who do not, or are unable to, access the Downs prefer not to fund it through their council taxes, which could be directed towards other services. On the other hand, local people who most benefit from personal use but are also affected by the nuisance of large events would prefer to minimise such activity by sustaining public funding. Measures to mitigate concerns on either side are evident in the committee's plans. The enhanced emphasis on education activities, for example, as discussed earlier in this chapter, could provide a means to broaden engagement across the city. A commitment to supply more detailed and timely information on events and the diversification of the programme to include more small-scale food festivals, charitable events, arts and well-being activities aims to assuage local residents' grievances.[145] Under Government constraints on public finances, however, the balance between cost and benefit and how it relates to different publics across the city of Bristol remains precarious. Arguments can and are made by local authorities on the role urban greens play in mitigating impacts of climate change, including high temperatures and local flooding, and their value for citizens' health and well-being. Yet, if city dwellers lack opportunities to engage both cognitively and affectively with the natural environments, they may be less likely to support pro-environment action, including when engaging with difficult decisions over where to direct funding.

Notes

1 Report of Law Officers regarding Durdham Down Enclosure, 4 May 1859, SMV/6/5/6/5, Bristol Archives.
2 'Scraps and Shreds', *Western Times*, 31 December 1842, p. 4.
3 Mark Bowden et al. (eds.), *An Archaeology of Town Commons in England: 'A very fair field indeed'* (Swindon: English Heritage, 2009), pp. 30–35; Richard Goldthorpe, *Clifton and Durdham Downs: A Landscape History Final Report* (Bristol: Bristol City Council, 2006), p. 2.
4 For historic descriptions of the site, see William Worcestre, *The Topography of Medieval Bristol* (1480), trans. Frances Neale (Bristol Record Society, 2000), p. 35, Section 56; Thomas John Lloyd Baker, 'An Account of a Chain of Ancient Fortresses, Extending Through the South Western Part of Gloucestershire', *Archaeologia* 19 (1818): 161–175; Goldthorpe, *Clifton and Durdham Downs*, p. 2.
5 Harry Mengden Scarth, 'The Camps on the River Avon at Clifton, with Remarks on the Structure of Ancient Ramparts', *Archaeologia* 44.2 (1872): 428–434.
6 J. Russell, 'The Archaeology of the Parish of Clifton', *Bristol Avon Archaeology* 16 (2000): 75.
7 SMR 20659. Goldthorpe, *Clifton and Durdham Downs*, p. 2; Russell, 'The archaeology of the parish of Clifton', pp. 73 and 83.
8 Goldthorpe, Clifton and Durdham Downs, p. 3.
9 Goldthorpe, Clifton and Durdham Downs, p. 3.
10 See John Latimer, 'Clifton in 1746', read at Clifton Antiquarian Club on 28 November 1900 and reproduced in the proceedings of the club Volume 5, pp. 25–34.
11 Latimer, 'Clifton in 1746', p. 29.
12 Amanda Cameron, 'Bristol Zoo Parking: Report Into Downs Licence and Legal Challenge Won'T Be Made Fully Public', *Bristol Post*, 12 July 2021. Available at: www.bristolpost.co.uk/news/bristol-news/bristol-zoo-parking-report-downs-5641486 (Accessed 10 October 2021).
13 John Latimer, *The History of the Society of Merchant Venturers of the City of Bristol. With some Account of the Anterior Merchants' Guilds* (Bristol: J. W. Arrowsmith, 1903), p. 167.
14 Celia Fiennes, *Through England on a Side Saddle in the Time of William and Mary* (London: Field and Tuer, The Leadenhall Press, 1888). https://visionofbritain.org.uk/travellers/Fiennes/26. For a similar description of the water, see Julius Ibbetson, John Laporte and John Hassell, *A Picturesque Guide to Bath, Bristol Hot-Wells, The River Avon, and the adjacent country; Illustrated with a set of views, taken in the summer of 1792* (London: Hookham and Carpenter, 1793), p. 168.
15 Latimer, 'Clifton in 1746', p. 25.
16 Latimer, 'Clifton in 1746', p. 27.
17 Ibbetson, Laporte and Hassell, *A Picturesque Guide*, pp. 197–198; John Latimer, *The Annals of Bristol* (Frome and London: Printed for the author [by Butler and Tanner], 1893), p. 266.
18 Latimer, 'Clifton in 1746', p. 26.
19 Goldthorpe, *Clifton and Durdham Downs*, p. 24.

20 Bristol Archives: SMV/6/5/6/1. Durdham Down Enclosure, 1 February 1859. This case is explored more fully in Chapter 6.
21 See Chapter 7.
22 Bristol Archives: SMV/6/5/6/9. New Drive, 1869.
23 Ibid.
24 'The Downs Committee', *Western Daily Press*, 31 July 1875, p. 6. Available at: www.britishnewspaperarchive.co.uk (Accessed 13 August 2020).
25 Bowden, Brown, and Smith, *An Archaeology of Town Commons in England*, p. 32.
26 *An Account of the Lead Mines producing Callamie, &c. On Durdham Downe, near Bristol, with a Proposal for the Disposing of a small Part thereof* (June 1714?) Bodleian Library, University of Oxford. As cited in Latimer, *The Annals of Bristol*, pp. 105–106.
27 Historic Collection (HC) Reference: 3583, 'lime kiln', 2011, *Know Your Place*, Bristol City Council, https://www.bristol.gov.uk/residents/planning-and-building-regulations/conservation-listed-buildings-and-the-historic-environment/historic-bristol/know-your-place
28 Ibid.
29 Pierce Egan, *Walks Through Bath*, p. 279.
30 John Latimer suggested that an annual horse race probably took place on Durdham Down prior to 1700. Latimer, *The Annals of Bristol*, p. 24.
31 Latimer, *The Annals of Bristol*, pp. 122 and 278–279.
32 Gerry Nichols, *Clifton and Durdham Downs: A Place of Public Resort and Recreation* (Bristol Branch of the Historical Association), p. 22; Rolinda Sharples, 'Horse Races on Durdham Down', City of Bristol Museum and Art Gallery.
33 *Salisbury and Winchester Journal*, December 20, 1824 (Issue 5395). Available at: www.britishnewspaperarchive.co.uk (Accessed 7 August 2020).
34 Nichols, *Clifton and Durdham Downs*, pp. 2 and 13.
35 Hammersley Plan. See also Donald Jones, *A History of Clifton* (Phillimore, 1992), pp. 12–13.
36 Nichols, *Clifton and Durdham Downs*, p. 1; Latimer, *The Annals of Bristol*, p. 23.
37 Bristol Archives: Society of Merchant Venturers (SMV) 6565 (1) and (2) 'Mr Pigott's Proposals for Windmill on Durdham Down'.
38 http://museums.bristol.gov.uk/details.php?irn=115404.
39 Goldthorpe, *Clifton and Durdham Downs*, p. 13.
40 'Meeting of the "Royal" on Durdham Down', *Harper's New Monthly Magazine*, Volume 58 (December 1878 to May 1879), p. 218.
41 Ibid.
42 *Western Daily Press*, 24 May 1937 (Issue 26372). Available at: www.britishnewspaperarchive.co.uk (Accessed 7 August 2020).
43 *Western Daily Press*, 15 May 1937 (Issue 26365). Available at: www.britishnewspaperarchive.co.uk (Accessed 7 August 2020).
44 *Western Daily Press*, 29 June 1936 (Issue 26093). Available at: www.britishnewspaperarchive.co.uk (Accessed 29 May 2020).
45 *Western Daily Press*, 3 August 1942 (Issue 27984). Available at: www.britishnewspaperarchive.co.uk (Accessed 20 July 2020).
46 *Western Daily Press*, 8 July 1946 (Issue 29200). Available at: www.britishnewspaperarchive.co.uk (Accessed 7 August 2020).
47 Oral History Interview (Bristol) 1("OHI Bristol" 1).
48 OHI Bristol 1.
49 OHI Bristol 1.
50 OHI Bristol 1.
51 OHI Bristol 2.
52 OHI Bristol 1.
53 OHI Bristol 1.

54 OHI Bristol 2.
55 OHI Bristol 3.
56 OHI Bristol 4.
57 OHI Bristol 4.
58 OHI Bristol 5.
59 Joshua Sprigg, *Anglia rediviva Englands recovery being the history of the motions, actions, and successes of the army under the immediate conduct of His Excellence Sr. Thomas Fairfax, Kt., Captain-General of all the Parliaments forces in England...* (London, 1647), p. 108.
60 HC: 3664, 'War time telegraph post', 2014, *Know Your Place,* Bristol City Council.
61 Latimer, *The Annals of Bristol*, pp. 137–138; 227 and 294–296.
62 Ibid., p. 83.
63 *Morning Post*, 24 September 1853 (Issue 24886), *Western Daily Press*, June 18, 1930 (Issue 24212), *Gloucestershire Echo*, 3 April 1934. Available at: www.britishnewspaperarchive.co.uk (Accessed 26 May 2020).
64 Section 10 Clifton and Durdham Downs (Bristol) Act 1861.
65 See the discussion of grazing on the Downs in Gerry Nichols, *To Keep Open and Unenclosed: The Management of Durdham Down Since 1861* (Bristol: Bristol Branch of the Historical Association, 2005), pp. 4–7.The evidence points to a maximum of 400 being depastured annually in the mid-Victorian period.
66 Bristol City Council (Commons Registration Authority), Register of Common Land, Register Unit CL0005, entry 1.
67 See Bristol City Council (Commons Registration Authority), Register of Common Land, Register Unit CL0001, entry 1.
68 See Gadsden and Cousins on Commons and Greens (3rd edition, Sweet & Maxwell, 2020) at paras 2.22–2.24.
69 Nichols, *To Keep Open and Unenclosed*, p. 7.
70 *A Management Plan for the Downs*, (Bristol City Council) para 2.6.
71 Clifton and Durdham Downs Bye-Laws 1998 and 2003, bye-law 7.
72 See Section 10, Clifton and Durdham Downs (Bristol) Act 1861.
73 Downs for People News Release, 21 January 2022, *The Downs are for People not for Cars*, para 3.
74 Nichols, *To Keep Open and Unenclosed*, p. 3.
75 Section 5 Clifton and Durdham Downs (Bristol) Act 1861.
76 See Nichols, *Clifton and Durdham Downs*, pp. 15–16.
77 Clifton and Durdham Downs Bye-Laws 1998 and 2003, bye-law 2.
78 Ibid., bye-laws 4 and 5.
79 Ibid., bye-laws 8 and 9.
80 Sections 16 and 17, Clifton and Durdham Downs (Bristol) Act 1861.
81 *A Management Plan for Clifton and Durdham Downs 2012–2017* (Parks and Estates: Bristol City Council, 2012). Available at: https://www.bristol.gov.uk/documents/20182/32863/The+Downs_v3.pdf
82 See *A Management Plan for Clifton and Durdham Downs 2012–2017* at pp 7 ff. for the plan's vision and objectives in more detail.
83 See section 7 of *A Management Plan for Clifton and Durdham Downs 2012–2017*, ibid.
84 See *A Management Plan for Clifton and Durdham Downs 2012–2017* at p 21.
85 Ibid., para 7.3.9
86 Ibid., para 7.4.
87 See ibid. at pp. 22, 23.
88 Originally designated an Area of Special Scientific Interest under the National Parks and Access to the Countryside Act 1949.
89 Avon Gorge SSSI notification 1003073.

90 Avon Gorge SSSI VAM 1003073.
91 Avon Gorge Woodlands Special Area of Conservation, Natura 2000 Standard Data Form UK0012734.
92 OHI Bristol 4.
93 OHI Bristol 3.
94 OHI Bristol 5.
95 OHI Bristol 3.
96 OHI Bristol 3.
97 OHI Bristol 3.
98 The research team carried out workshops with two school groups from Westbury Park School and a public engagement with the Friends of the Downs and Avon Gorge. For further details on the participatory research activities carried out for this project, see Chapter 9.
99 OHI Bristol 1.
100 Danielle Bashan, Agathe Colleony, and Assaf Shwartz, 'Urban Versus Rural? The Effects of Residential Status on Species Identification Skills and Connection to Nature', *People and Nature* 3 (2021): 347–358.
101 Walking Conversation with Tim on 27 February 2022.
102 Ibid.
103 Ibid.
104 See Julie Whitburn, Wayne Linklater and Wokje Abrahamse, 'Meta-Analysis of Human Connection to Nature and Pro-environmental Behaviour', *Conservation Biology* 34:1 (2019): 180–193.
105 See *Downs Committee Survey, Principles and Workplan, Final Results* (The Downs Committee, 2022), p. 16. Available at: https://democracy.bristol.gov.uk/documents/s72559/Appendix1-Fullsurveyresults.pdf
106 Ibid., p. 22.
107 OHI Bristol 1.
108 OHI Bristol 1.
109 See *Downs Committee Survey. Principles and Workplan. Final Results* at p. 22.
110 Ibid., p. 16.
111 See *Clifton Down Wildflower Meadow Proposal* (The Downs Committee, 2022). Available at: https://democracy.bristol.gov.uk/documents/b31033/CliftonDownWildflowerMeadowProposal19th-Oct-202214.00DownsCommittee.pdf?T=9.
112 Section 10 Clifton and Durdham Downs (Bristol) Act 1861.
113 See *Downs Committee Principles* (The Downs Committee, 2022), p. 1. Available at: https://www.bristol.gov.uk/files/documents/4822-downs-committee-principles/file
114 OHI Bristol 1.
115 Walking Conversation with Tim on 27 February 2022.
116 See Jody Aked, Nic Marks, Corrina Cordon, and Sam Thompson, *Five ways to Wellbeing* (London: New Economics Foundation, 2008), pp. 1–23.
117 Ibid., p. 3.
118 For further examples of applications of the Five Ways to Well-Being, see Jody Aked, Nic Marks, Corrina Cordon, and Sam Thompson. *Five Ways to Wellbeing: New Applications, New Ways of Thinking* (London: New Economics Foundation, 2011).
119 Walking Conversation with Tim on 27 February 2022.
120 Ibid.
121 Ibid.
122 Natalie Smith, *In-Common Sites Evaluation Form* (Wastes and Strays, 2022).
123 Ibid.
124 Walking Conversation with Tim on 27 February 2022.
125 OHI Bristol 1.

126 See Parks and Green Spaces – future funding model 2018–2020 (Parks and Estates: Bristol City Council, 2018), pp. 1–13. Available at: https://democracy.bristol.gov.uk/documents/s21388/Appendix20A2020ParksGreenSpacessavingscabinetreport.pdf
127 Ibid., pp. 2, 3, and 5.
128 See earlier in this Chapter, p. 91–92.
129 Bristol Pride is a donation entry rather than a ticketed event.
130 Open Spaces Society (2021). *The Downs are for people, not for cars.* Available at: https://www.oss.org.uk/the-downs-are-for-people-not-for-cars/
131 Suzanne Audrey, *Public Statement to the Downs Committee, 5 July 2021* (The Downs Committee, 2021), para 2 & 7. Available at: https://democracy.bristol.gov.uk/documents/b28017/PublicForumandTermsofReference05th-Jul-202112.00DownsCommittee.pdf?T=9
132 Ibid., para 7.
133 Downs for People. *Public Forum Statement: Downs Committee Review* (The Downs Committee, 2021), para 1. Available at: https://democracy.bristol.gov.uk/documents/b28017/PublicForumandTermsofReference05th-Jul-202112.00DownsCommittee.pdf?T=9
134 See *Downs Committee Principles.* Available at: https://www.bristol.gov.uk/files/documents/4822-downs-committee-principles/file
135 See *Downs Committee Work Plan 2022–23* (The Downs Committee, 2022). Available at: https://www.bristol.gov.uk/files/documents/4821-downs-committee-work-plan/file
136 See *Downs Committee Survey. Principles and Workplan. Final Results*. Available at: https://democracy.bristol.gov.uk/documents/s72559/Appendix1-Fullsurveyresults.pdf
137 See *Downs Committee Principles.*
138 See *Downs Committee Workplan 2022–23*, p. 2.
139 See *Downs Committee Survey. Principles and Workplan. Final Results*, p. 32.
140 Alex Seabrook *New Café Plans for the Downs Would Likely be Blocked by 160-Year-Old Victorian Law* (Bristol Live, 1 March 2023).
141 Downs for People Public Statement, 13 October 2021, *Downs Review paper for meeting, 20 October 2021*, point 13, para 4.
142 Ibid.
143 See *Downs Committee Survey. Principles and Workplan. Final Results*, p. 21.
144 Ibid., p. 21.
145 Ibid., pp. 25 & 27.

5 Valley Gardens, Brighton

Introduction

Location

The Steine, Valley Gardens and the Level are a series of green spaces, totalling c.92 acres, that collectively form the Valley Gardens Conservation Area under the administration of Brighton & Hove City Council (Figure 5.1). The green spaces run up the Wellesbourne valley from the Steine at Brighton's seafront to the Level, approximately one mile to the north. They connect the old town of Brighthelmstone to the west of the green space with the Regency development to the east. From the early nineteenth century, these green spaces formed an important junction for major traffic routes into Brighton, most notably London Road from the north and Lewes Road from the east. The area includes several listed buildings, most notably the Royal Pavilion – residence of the Prince of Wales, later George IV. Brighton's town authorities purchased the Pavilion in 1849–1850, ensuring the building's protection as well as that of the surrounding green spaces. The conservation area was subject to a multimillion pound investment project from 2018 to improve the road network whilst enhancing the publicly accessible open space.

Topography/Features

Historically, the Valley Gardens area formed a drainage system for the Wellesbourne that ran from springs on the chalk downs at Patcham, down the length of the green space, and entered the sea at Pool Valley. Prior to 1793, overflow from the Wellesbourne formed a large pond on the Steine. A geological survey conducted in 1864 shows the course of the Wellesbourne as a seam of London clay amid the chalk. The Prince of Wales and Duke of Marlborough funded construction of a new sewer in 1793, but the issue of flooding was only partially resolved with episodic flooding from rainwater runoff continuing to be an issue. The southern part of the Steine was fenced off, landscaped and planted in the early nineteenth century. From 2020, Valley Gardens has a newly created landscaped green space with architecturally designed planting, pathways and open public spaces that are surrounded by a busy road network. As such, neither the planting schemes undertaken in the 1820s nor the original open grassland have survived.

DOI: 10.4324/9781003204558-6

Figure 5.1 Map of Valley Gardens, Brighton.

The combination of seasonal flooding and urban development has limited archaeological survival. There are some indications of Romano-British deposits that suggest human interaction with the green space before it was in use as a common. But the material evidence for use of the site as an open space is limited, especially for the period prior to the onset of urbanisation in the late eighteenth century. The shift in focus to recreational uses at this time impacted the character of the green space and contributed to landscape change, but also continues to add to the site's historic appeal by way of important built features, such as the Pavilion, St Peter's Church and the Victoria Fountain.

Narrative

The Domesday Book reference to Brighthelmstone in 1086 is the earliest detailed evidence of a fishing settlement at present-day Brighton, although given regular erosion by the sea, it is possible that the whole settlement was remodelled more than once. A painting from c. 1545 showing an earlier naval attack is the earliest surviving visualisation of the settlement (Figure 5.2).

It shows Brighton's rectangular street layout bounded by the open space of the Wellesbourne valley on its eastern edge and surrounded by arable land. Conflicts between fishermen and "landsmen" over contributions to the upkeep and defence of the town were investigated by commissioners on behalf of Elizabeth I, resulting in the drafting of *Elizabethan Brighton: The Ancient Customs of Brighthelmston 1580*. The commissioners also established the Society of the Twelve comprising "the auncientest gravest and wysest inhabitants eight ffishermen and fower landemen"[1]. to assist the town's officials and make decisions on behalf of the residents.[2]

The early modern period witnessed Brighton's transformation from a fishing village to a fashionable seaside resort, and this shift impacted directly on the Wellesbourne valley. From the sixteenth century, the area was part of the freehold property of the lords of the manor of Brighthelmstone, but the green space was viewed by the local community as a public open space and was used for a range of activities. The town experienced economic decline in the early eighteenth century, and by 1750, Brighton was a poor seaside town with a population of about 2,500.[3] The area was rejuvenated in the later part of the century, thanks to local doctor Richard Russell's advocacy of sea bathing and seawater drinking, which led to the town's reinvention as a resort.[4] After Russell bought property at the southern end of the Steine in 1753–1754 to accommodate patients taking his cure, this area of green space became a natural focal point for the associated regeneration and an important location for elite social interaction. The space thus contributed to Brighton's economy as a site for recreational pursuits, including through the establishment of libraries in the immediate vicinity.

In 1784, the Prince of Wales leased a house on the Steine and it was at this point that Brighton became synonymous with royalty. By the late eighteenth century, Valley Gardens had become a place of summer retreat and entertainment for the

Figure 5.2 Naval Assault on Brighton in 1514, 1545.

Prince and his friends, but also for members of the gentry and professional classes. As Brighton's popularity increased, so too did the built environment, with property construction occurring along the eastern and western edges of the green space. In April 1822, the Commissioners of Brighton negotiated with the lords of the manor and the green space was vested to trustees who acted on behalf of Brighton's residents. The powers of the trustees were then formalised within the role of the borough council under the terms of the 1873 Brighton Borough Extension Act.

Governance and Property Rights – Historic

The Changing Space of the Common

The transformation of Valley Gardens was a continuous process throughout the eighteenth and nineteenth centuries. Initially a completely open space, it was gradually converted into distinct and bounded landscape units as a direct result of urbanisation.

In the eighteenth century, the green space was sufficiently open to the South Downs that the Duke of Cumberland turned out stag two or three times a season, which he would hunt as a spectacle for the population of Brighton. The space was also used for grazing animals, and the area at the southern end of the Steine was commandeered by the fishing community for activities associated with their work. However, the space began to change following increased tourism in the mid-to-late eighteenth century.

Enclosure and "Improvement"

A map of Brighton from 1792 (Figure 5.3) shows the Steine being partially enclosed from the greater extent of the green space. During this first phase, wooden railings were used, though these were replaced with iron ones in 1823. The innovation was designed to allow only foot traffic to pass onto the Steine, thus protecting it from earlier uses associated with agriculture or fishing; although, if paintings of the late eighteenth and early nineteenth centuries are to be believed (for example Figure 5.4), some activities like boat storage and sheep grazing initially continued alongside the new recreational uses.

The "improvements" and expansion of Brighton from the 1790s further impacted the green space. In 1793, the Prince of Wales and the Duke of Marlborough, who both owned houses in the vicinity, offered to undertake improvements to drain the water from the Wellesbourne that collected as a lake in front of their properties. The improvements consisted of an arched sewer running across the Steine with two entrances – one at the centre of Pavilion Parade and the other at the east entrance of Castle Square (now the intersection of Old Steine and St James's Street). The sewer discharged into the sea at the back of William's Baths, which was located on the south side of the Steine (close to the Albion Hotel). The cost of the new sewer was met by the Prince and Duke. In return, the lords of the manor of Brighthelmstone – with the consent of the Society of the Twelve – granted permission to the Prince and Marlborough to rail (thus enclosing) a portion of the open space that adjoined

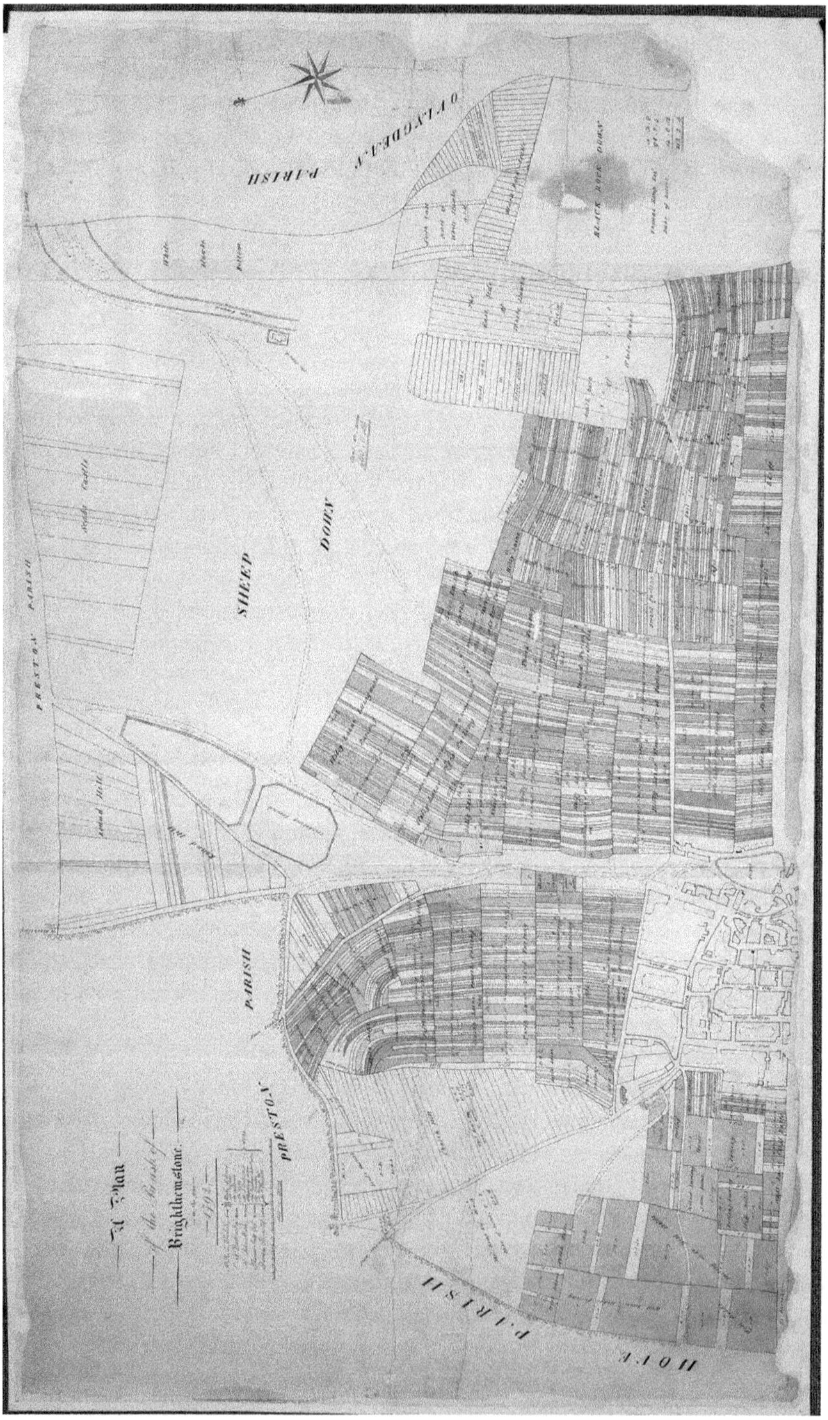

Figure 5.3 A Plan of the Parish of Brighthelmstone in the year 1792.

Figure 5.4 Brighton. A Bird's-Eye View from the Preston Road, 1819.

their properties, creating the gardens that now form the Eastern Lawns. The agreement included the condition that they could never "build or encumber with anything that might obstruct the prospect or be in any shape or way a nuisance".[5]

The resulting enclosure of this previously open green space to the north of the railed Steine essentially created a private garden to the front of the Pavilion, albeit with low railings that allowed visual continuity across the space. Initially, the enclosures in front of the Pavilion and Marlborough's property would have been seamlessly integrated with the surrounding green space because the road network only consisted of informal footpaths or trackways. These changes were, however, part of the gradual transformation of the Steine. While it was not formally landscaped until the 1820s, the dominant use of the green space was already coming to be more strongly associated with recreation and the needs of tourists rather than the earlier grassland activities. These initial enclosures, coupled with the changing use, created a domino effect which saw the gradual enclosure and landscaping of the open space up to the Level.

Further changes were made at the behest of Prince George in the early nineteenth century. In 1803, he sought permission to enclose the London Road that ran behind the Pavilion. This permission was granted in 1805 at the Lewes Session, but with the following clause:

> should the Prince at any time cease to require the use of the same, proposed to be given up by the inhabitants, or should the Pavilion and grounds at any time pass from his occupation or be sold, then and in either of the said cases the said portion was to revert to the use of the inhabitants.[6]

Furthermore, in 1811, the Prince purchased Promenade Grove, which was a popular pleasure garden to the west of the Marine Pavilion. This purchase and subsequent enclosure was also granted on the condition that he fund the building of a new road. These enclosures forced traffic in front of the Pavilion, which had lasting consequences, with the green space increasingly encroached upon by roads. As existing footpaths and trackways were formalised into street systems using sand, shells and gravel, the green space became divided. In place of the original open grassland, there emerged smaller enclosed grassed areas separated from each other by roads, paths or railings. However, the 1805 clause also provided the legal means to save the Pavilion for the town in 1849.

The Pavilion

The Pavilion itself also impacted on the aesthetics of the space. The original building, which was described as a "superior farmhouse", was extended by Henry Holland in 1787 with the addition of a circular saloon and an extra wing. Further additions followed, culminating in the complete redesign of the exterior in Indian style by John Nash between 1815 and 1823. Such a striking building inevitably affected the feel of the open space in which it stood.

The Impact of Regency Expansion

Change continued as the nineteenth century progressed. Between 1808 and 1830, at least four new east to west streets were constructed, resulting in the enclosure of the green space up to and including the Level, and in 1834, a further road cut across the southern part of the Steine to give direct access from eastern Brighton to the town centre. This process was partially a result of the expansion of Brighton's built environment during the Regency period. As the fields to the east and then north of the green space were converted into streets and housing, Valley Gardens became a necessary communications link between the old and new sections of the settlement.

At the same time, the possibility of expanding Brighton's residential space arose in part from the attraction and economic exploitation of the green space. This too was advanced in this period with the appointment of the first landscape managers for the Steine and Level (seven in total between 1822 and 1824). It was their responsibility to lay out walks and planting schemes designed by Amon Henry Wilds and Henry Phillipps. It was these planting designs, in conjunction with the creation of St Peter's Church (1824–1828), which fully realised the characterisation of the space as green traffic islands running from the beachfront to the Level.

Perhaps the most remarkable feature of Valley Gardens is the preservation of its boundary, despite the numerous changes that have occurred within. The expansion of Regency Brighton took over the surrounding agricultural land rather than the green space itself. Consequently, if the roads were to be removed, the open unenclosed green space could be reinstated relatively easily. In the twenty-first century, the nature of the space as a series of traffic islands has presented multiple infrastructural and design problems. In 2018, a phased work package began on a multimillion pound project to redevelop Valley Gardens as a new central park running from St Peter's Church to the seafront. Far from providing the infrastructural cohesion intended by the nineteenth-century developments, the green space has become cut off from the community surrounding it and has been negatively impacted by increased traffic flow.

The Urban Common in Social and Political Context

The reinvention of the Wellesbourne valley from open grassland to a series of green traffic islands reflects Brighton's transformation from a fishing village to a seaside resort. These shifts simultaneously drove – and were driven by – the changing use of the space.

Early Activity

The paucity of archaeological material from the Wellesbourne valley makes the assessment of the pre-modern history of the space difficult. There is some evidence of Stone and Bronze Age activity in the form of implements and of a Roman presence reflected in burials and coin hoards. Two nineteenth-century antiquarian accounts connect the Steine to Roman occupation. One noted the place name evidence Steyne or Stone street suggesting the presence of a Roman road in the vicinity. The other claimed that a Roman camp was located on the Steine.[7] More recently, Ken Fines has highlighted the evidence for a small port at the mouth of the Wellesbourne in the form of foundations buried in the seabed south of Palace Pier.[8] The coin finds in the area may be related to votive offerings at this spot.

Fishing Settlement

A fishing settlement was located at Brighthelmstone at least from the eleventh century, as recorded in the Domesday Book.[9] During the sixteenth and seventeenth centuries, Brighton's fishing fleet was one of the most important along the south coast. By 1579, there were said to be 80 fishing boats operating out of the town employing 400 mariners who collectively owned 10,000 nets.[10] There was a history of jealousy and strife between the fishermen and the landsmen, and in 1580, a commission was sent to the town to settle the differences over payments for the upkeep of public amenities and the defence of the town.[11]

The Steine was used by the fishing community for drying nets, to store timber and boats and for boat construction and repair. There is also evidence that the Steine came to be used as a more general workspace for local residents. Court rolls from the time of Elizabeth I state that no hog could go un-ringed on the Steine "where nets lie" with a financial penalty for any breach.Carriages, such as timber vehicles, waggons and carts, were brought onto the space for repair.[12] There is also reference to coal storage during the eighteenth century and to its use as a sales area by local merchants.[13]

Eighteenth-Century Transformation

The main uses to which the open space of the Wellesbourne valley was put changed dramatically with the transformation of Brighton's fortunes in the eighteenth century. The early part of the century saw economic depression resulting from a decline in the local fishing industry arising from a series of bad storms. Its fortunes were soon arrested, however, and by 1800, Brighton was said to be "the most frequented [and] without exception one of the most fashionable towns in the Kingdom".[14] The area became a focus for health tourism and recreation, prompted initially by the recommendations of Richard Russell regarding the benefits of seawater bathing and drinking and then through the kudos brought to the area from its patronage by the Prince of Wales.

Royal Patronage

Prince George first visited Brighton in the summer of 1783, aged 21, when he stayed with his uncle the Duke of Cumberland who had bought the property Dr Russell had owned at the southern end of the Steine. The following year, the Prince returned, renting a house of his own on the Steine. In 1786, his German cook and factotum Louis Weltje arranged for him to rent what was described at the time as a "superior farmhouse". Between that point and 1823, this building was transformed into the striking Pavilion that still exists today. After becoming King in 1820, George IV spent less time in Brighton, but its reputation was well established by then and royal patronage continued. William IV responded to a loyal address from the town, saying: "Tell the inhabitants of Brighton that I shall soon be with them".[15] He travelled down less than two months after becoming king, following this a couple of weeks later with his official entry alongside his wife Queen Adelaide. After this, the royal couple regularly entertained at the Pavilion. Queen Victoria visited Brighton in October 1837, and early in her reign held musical evenings at the Pavilion. However, she and Prince Albert were less enthusiastic about the town, particularly after the arrival of the railway brought large numbers of new holidaymakers. She complained about the lack of privacy: "The people are very indiscreet and troublesome here, which makes this place quite a prison". The Pavilion was abandoned by the royals in 1845, and four years later, it was purchased by the authorities for the inhabitants of the town.

Despite not being particularly fond of Brighton, Queen Victoria did exert an imprint on the Wellesbourne valley. The Victoria fountain was built at the centre of the Steine gardens in 1846, as a somewhat tardy commemoration of the coronation almost a decade earlier. Sir John Cordy Burrows (who was later Mayor of Brighton) was disappointed that the occasion had not been formally marked and mounted his own subscription campaign to fund the fountain, which was designed by the architect Amon Henry Wilds. Burrows was also responsible for laying out the surrounding gardens. In a further twist, the fountain quickly became a platform for political oratory, the Steine in general having first been used as an arena for this purpose around the time of the Reform Bill in the 1830s.[16] The legacy of this activity continues today, with the Level remaining a popular site for demonstrations, rallies and protests.

Recreational Activities

Two key activities that took place on the open space of the valley in the late eighteenth and early nineteenth centuries were promenading and people watching. Indeed, the Steine effectively became a free pleasure garden that was paid for and managed by the town commissioners and was accessible to residents and visitors alike.[17] The Steine appears to have been a place for walking as far back as the 1760s and also for skating in severe weather when pooled water froze.[18] There are several newspaper reports and private diary entries quoted by John Erredge that date from the early nineteenth century referring to the Prince of Wales and members of the nobility walking, riding on horseback or circling the space in a carriage to the delight of the promenading public.[19] Developments to facilitate these activities included the laying of a brick promenade around the southern end of the Steine (inside the railings) in 1806, the landscaping and planting of the 1820s and the decision in 1824 to light the southern end of the Steine.

Organised sports also took place in the Valley Gardens area. The *Morning Herald* for 11 September 1805 described several races that took place on the Level:

> A pony race on the Level, this morning afforded much diversion to a very numerous assemblage of spectators. After this, donkey races took place: seven started for the first heat, and what is very singular, two, on this starting, ran a dead heat; a circumstance, probably, with quadrupeds of this sluggish tribe, never recorded in the annals of sporting. The donkies having performed their task, the company removed to the Steyne, to the South, where jumping in sacks, and a jingling match kept hilarity alive for about two hours longer.[20]

As the Prince Regent was fond of cricket, the Lord of the Manor Thomas Kemp granted an area to be used for the purpose. This practice seems to have continued even after the Prince's death, since an image from 1849 shows a cricket match being played on the Level between a Sussex team and one from Kent.[21] In 1822, Mr James Ireland purchased an area to the north of the Level and established a public garden alongside the cricket ground with a bowling green, billiard room and a colonnade with seats.

Alongside sporting activities, betting also took place on the Steine. An account from 1792 refers to "Sporting men of fashion, dashers, and blacklegs" assembling on the Steine to make bets for the Lewes races.[22] The space was also used for other recreational gatherings and events. In 1805, a marquee was erected on the Steine in order to exhibit a monster starfish that had been found, and between 1807 and 1811, an annual sheep fair was held which included gingerbread stalls and roundabout rides.[23]

Recreational activities within the space were encouraged by the provision of certain facilities and the construction of specific buildings around the site. From the 1750s, bathing machines were made available, Assembly Rooms were opened in Ship Street (1767)[24] and a tower was constructed from where musicians could entertain promenaders at the southern end of the Steine.[25] Captain Wade was appointed as Master of Ceremonies to organise events for visitors.[26] This reinvention as a public pleasure garden also saw new regulations imposed on the space. Running, racing, cricket, trap ball and other games were all prohibited (except in designated areas) and offenders were subject to prosecution.

Particularly popular in the eighteenth and nineteenth centuries were the subscription libraries. The first was said to have been opened by Mr Baker in 1760 and was located on the east side of the Steine. A rival premises was opened at the south end by Mr Woodgate in 1767; it was one of several that proved popular in the late eighteenth century. By the early nineteenth century, the leading institutions were Lucombe's (originally Donaldson's Library) and Turner's. In 1827, the annual subscription fee at Lucombe's was £1. 11s. 6d.[27] The libraries were designed so as to make the most of the open space of the Steine with many having balconies and outdoor seating. They were often used as entertainment spaces with music and drinking on offer alongside the books.

Military Uses

Military bands sometimes formed part of the recreational entertainment, but the space was also used for more serious military purposes, including the storage of weapons and for military camps and training. A map from 1779 shows a battery at the bottom of East Street and there were reports of eight guns being deposited on the Steine for several weeks. Militia camps appeared at several points in the late eighteenth century and a diary entry from 1805 refers to members of the nobility (including royalty) training on the Steine. During the conflicts of the early twentieth century, the area – like many other open spaces – was subject to bomb damage.

Contemporary Governance and Property Rights

As is the case with all of our case studies of urban commons, the contemporary governance of Valley Gardens has been determined and shaped by its history and the manner in which the different uses to which the land has been put over the centuries – by both the community and individual landowners (in this case, including the Prince of Wales) – has been captured in property "rights" with normative legal force. In terms of modern governance, however, the similarities end there. The contemporary governance arrangements for Valley Gardens differ markedly from those of the other case studies. The land emerged from the manorial system at a much earlier date, as noted above. Although there was some agricultural use on parts of what is today Valley Gardens, this almost ceased by the end of the eighteenth century and was not reflected in established property rights giving rights to graze livestock or take estovers or turbary. None of the land was registered as common land under the Commons Registration Act 1965, and there are no subsisting common rights over it. There are no statutory designations for the protection of wildlife or landscape, unlike those (such as Sites of Special Scientific Interest [SSSIs]) in some of our other case studies. And, perhaps most important of all, the common is not statutorily protected by a bespoke Act of Parliament of the kind that protects Town Moor (Newcastle) or Mousehold Heath (Norwich) from development. The "common" is instead dependent for its protected status on spatial planning policy, in particular measures introduced by Brighton City Council through the town and country planning system. While this may be seen as a weakness, and in the past has led perhaps to a poorer level of protection for the green space provided by Valley Gardens, it has also proved to be a strength: it has enabled an imaginative and forward-looking redesign of the space to be implemented without the constraints that would otherwise be imposed by the complex property rights and statutory requirements that apply to many other urban commons.

Valley gardens is designated as a Conservation Area within Brighton's spatial development plan. A detailed study of the conservation area, including problems and suggestions for future policy to promote and protect the Gardens as a green space for the community, was conducted in 1995.[28] The assessment noted that the area is not a homogenous land unit, with different parts having developed differently historically and with differences in land use, character and appearance. It therefore presented an assessment – and option for future development and use of the green space – based on four sub-areas: the Old Steine/seafront, Royal Pavilion and New Road, Victoria Gardens (the central position of the case study) and The Level at its northern end. The unique character of Valley Gardens overall was stressed to be its role as a "green corridor" of open space in the urban centre of Brighton – and a "transition space" between the characters of the urban developed space on the eastern and western sides of the Gardens.[29]. Among problems highlighted[30] were the encroachment of unsuitable building development on parts of the common space, the isolation and underuse of the central Victoria Gardens and Old Steine Gardens section of the case study and the heavy volumes of traffic on the roads running along and traversing the space – for example, in the case of Victoria Gardens, rendering the gardens "little more than traffic islands (albeit attractive ones)".[31] Among policies proposed to enhance the appearance and utility of the space were the increased use of article 4 directions to remove permitted development rights for minor alterations and additions to existing buildings in several parts of the case study[32] and the promotion of planning policies to protect the longer views of key historical elements such as the Pavilion, Dome and St Peter's Church.[33] A Conservation Area Partnership Scheme was proposed with English Heritage to promote many of the recommendations made in the study.[34]

Two further developments underpin the reshaping and "resurrection" of Valley Gardens as a vibrant public green space for Brighton. The City Council commissioned a CIVITAS-funded Public Realm Analysis in 2011, which highlighted the social and emotional benefits of an imaginative street design.[35] The Brighton and Lewes Downs Biosphere Reserve was dedicated by UNESCO in 2014.[36] Valley Gardens is centrally situated in the "transition" zone within the biosphere. Both contributed to the impetus for developing an innovative and imaginative vision for redesigning Valley Gardens to provide improved social benefits for residents, improved biodiversity in an urban setting and improving traffic management within the urban centre of the city.

The contemporary reshaping of the common space is being delivered through a three-phase urban development project – a Vision for Valley Gardens.[37] Phases 1 and 2 of the project redeveloped the area from St Peter's Church to Edward Street and was completed in 2020. Phase 3 (Old Steine to Palace Pier) was approved in 2019 and commences in 2023. Central government made eight million pounds available for phases 1 and 2 through the Coast to Capital Local Enterprise Partnership (LEP) and a further six million pounds was awarded through the LEP for phase 3 in 2019. The scheme implements many unique and innovative features, including a 650 metres long river of flowers perennial garden that "flows" along the eastern side of Valley Gardens. The redesigned space hosts the National Elm Collection and is based on an imaginative landscape plan using the principles of flow from grey to green, hard to soft, roads to paths, lawns, gardens and trees.[38] It includes a new setting and square for St Peter's church, wildflower meadows, new gardens and extensive tree planting, combined with highway modifications to improve traffic flow through the city centre to the seafront, without adversely impacting the new linked park system implemented for the redesigned urban green space.

Present Voices, Future Directions

The development of Valley Gardens rooted discussions about the future into the present in a way that was perhaps less explicit in our other case study commons. Observations from local users reflected the recent changes in landscaping and infrastructure, the walking along the river of wildflowers, for example, or a perceived invitation to "linger" that was not evident before the renovation of the gardens. The understanding of urban green spaces as places to meet took on a new meaning during the COVID-19 pandemic. Through various phases of lockdown restrictions, people located their comings together in open green space: walking and talking in pairs, then, bubbles, expanded to lingering in groups of six, for conversation and the sharing of food, perhaps. There were also gatherings that pushed against the government's lockdown restrictions, following the tradition of urban commons as sites of protest and relative lawlessness. Restrictions on collective comings together also effected the capacity for public users to engage in this research project. Participatory research in Brighton's Valley Gardens encompassed self-directed creative explorations carried out by community participants and walking conversations with key members of the team involved in the regeneration project for Brighton and Hove City Council. In addition, groups of students from both the Universities of Portsmouth and of Brighton carried out design-based investigation to, in the words of tutor Graham Perring:

> explor[e] ways in which infrastructural changes can provide a catalyst for re-imagining the design of public realm architecture in sensitive locations

> [... and] expand their [students'] thinking to engage with wider concerns in relation to the nature and extent of the contemporary commons and its potential to exist as an expanded arena of community, operating beyond traditional, physical boundaries.[39]

As lockdown measures started to lift in the summer of 2020, citizens of Brighton came out to discover the newly reshaped Valley Gardens as phase two of the redevelopment plan completed.[40] Transformative and contemporary, the development of Valley Gardens brought future visions into a material present. So, a useful starting point for thinking about the future of urban green space here is a reflection on how the regeneration project came into being and what lessons might be learnt from its implementation. Valley Gardens, prior to the development, seemed to have been uninviting, tired and barely used. Jim Mayor, the project manager tasked by Brighton and Hove City Council with leading the regeneration of Valley Gardens, recalled that at that time it:

> ...was [not] really a place for anybody, particularly. It was a route through the city, it was a boundary, [...] had to be a boundary because of the traffic infrastructure. The guardrail really cut off the east and west sides of the city'[41]

Defined by its river valley geography and its history of enclosures, described earlier in this chapter, Valley Gardens acts as both boundary and conduit between the eastern and western parts of the city. This rationalisation of the Gardens as a place of transit has been compounded by the way policymakers tend to perceive open spaces in cities "under the bracket of transport spaces".[42] As Jim Mayor observed:

> Public spaces don't have a place in councils because of the transport-centric approach to design of the spaces between buildings"[43]

In the case of Valley Gardens, then, focus lay in the flow of vehicle traffic around it and pedestrian traffic across it, with little attention paid to the pleasure gardens themselves. As a consequence, there was little invitation for public users to stop and settle:

> Unless perhaps you're going to St Peter's there wasn't really very much else in the gardens to attract people and a lot of things to prevent people from wanting to cross the road to get into these things. Including lots of signs once you did get here that said, 'Stay off the grass'.[44]

In his reflection, Jim Mayor emphasised how the lack of appeal not only resided in the tired and perhaps outdated ornamental gardens but in regulatory measures, barriers that effectively restricted access and signage that curtailed modes of inhabitation once inside. The idea that "nobody used" the gardens was echoed by Emma Friedlander-Collins, who used to visit the ornamental gardens as a student 20 years ago:

> You could feel kind of quite tucked away and every now and then you'd see the odd drug dealer or someone in their pyjamas walking their dog. Like, that would be the only people you would see in that space at all.[45]

The absence of others in Emma Friedlander-Collins's recollection seemed to evoke both a sense of escape for the student newly arrived in the city and perhaps curiosity at the unfamiliar occasional visitor. Emma Friedlander-Collins also recalled, however, the "massive empty space" of The Level as "an intimidating place to go", where "you would have to go in groups of three" to walk across "because it was quite scary".[46] Other areas of Valley Gardens, "the abandoned boating area" and "particularly around St Peter's Church", were "used by street drinkers and people just wanting somewhere to hang out".[47] In her description, however, Emma Friedlander-Collins drew a distinction between the openness of The Level and the bounded character of these two sites. Moreover, as an urban planner, she perceived a correlation between the barriers and the creation of these uninviting spaces:

> So, they put the fences in to protect them. They basically made a pen, a street drinkers pen, which nobody else wanted to go into. You know, it's an incredibly intimidating space. And there would be 20 or so people in there, not doing any harm but just being really drunk and really loud and aggressive, so actually it became a really, a real deterrent.[48]

Where the guard rails were installed to prevent a street drinker inadvertently entering the road to become victim of a traffic accident, the containing fences also collected and confined people together, generating an inhospitable environment. Moreover, the transport-centric approach of separating people and vehicles for safety and quite probably to ensure traffic flow resulted, at St Peter's, in eight barriers between the green and other side of the street.[49] Consequently, the public, here, neither lingered within the gardens nor walked through to access the two sides of the city.

Following the 1995 study of this conservation area,[50] Brighton and Hove City Council sought funding for the regeneration of Valley Gardens, which once secured, led Jim Mayor and Emma Friedlander-Collins to initiate a public consultation. As an urban planner tasked with designing the future of the gardens, Emma Friedlander-Collins emphasised the importance of public consultation; planning developments should not simply be about saying "this is what you're having", but rather about facilitating "community engagement so that you're creating a space for the needs of the local community".[51] The challenge for the Valley Gardens consultation, however, was "trying to find community", and despite a year-long effort to engage stakeholders, it seemed as though "no one else really cared".[52] In contrast to our other case study commons, where communities articulated a sense of belonging, in Valley Gardens it seemed absent. The absence of community belonging may also be grounded in a lack of awareness of the historical antecedents of the Gardens as common land, which have been seen to exert powerful effect elsewhere. Another factor is evident in the descriptions above 'in the gardens' impoverished appeal, for use generates connection. Interestingly, however, Emma Friedlander-Collins also identified the challenge of enabling people to envision the future of the gardens. "They couldn't see what it was going to become", which impacted on people's motivation to "invest time and energy" attending meetings to "talk about something that was just scrap space".[53] As discussed in Chapter 4, city dwellers' relational encounters with green space, cognitive and emotional, underpins their engagement in environmental action and policy. Where there is a paucity of such interaction, so too this may impact on their ability to conceive or care.

In response to this dilemma, a different kind of "bottom-up" approach was adopted by the regeneration team in 2007, with the council first removing some of the enclosures "to try and get some initial [...] quick wins by taking out some of the unnecessary guard railing".[54] The intervention generated a profound effect on the use of the space, as Emma Friedlander-Collins recalled, "taking those fences down democratised that space very, very quickly. And because we did that successfully, we were then able to start taking down other fences".[55] Removing the enclosing railings was only part of the process of re-commoning Valley Gardens. Another simple intervention, the siting of "two massive benches [made] out of natural wood", also diversified usage and users:

> So, we put in two benches, took the fences down, and what was interesting was within a week there was mum picnicking in there with her child, there were people just sitting on the benches having coffee, and the street drinkers weren't an issue because they were balanced by other people in the space.[56]

Her reflection emphasised how the temporary changes in the Valley Garden environment prompted passers-by to interact in new ways, with the space and with each other. In this way, the interventions operated as a material form of public consultation, as urban planners observed how the public engaged with the various "invitations" to inhabit the space differently. One bench incorporated audio speakers to test out users' reactions to different sounds – trickling water, bird song, a radio play. The nuanced consultation informed decisions, such as emphasising tree planting to increase the bird population and, consequently, enhance birdsong in the gardens. Other invitations involved pop-up events – a week-long street market, a play bus, a community group building a temporary garden – aimed to generate audiences for more traditional forms of asking what people wanted in the space.

In tandem with this new community consciousness was the beginning of a different relationship with the Gardens as a natural location, and with that a desire to be "protective of this space"[57] in particular, a sense that Valley Gardens had a future that was inclusive and biodiverse:

> One of the ways that narrative could be continued into the future would be dealing with issues around climate change, making [it] more people-friendly, sustainable cities. So, there's an element of the… the planting contributing to that aspiration, biodiversity, environmental."[58]

The planting, Jim Mayor referenced here, is one of the most striking elements of the gardens' renovation, both visually and in the unseen complexity of its design. Where public users appreciated the vibrancy of the flow of flowers, the vision encompassed a more holistic approach to horticultural planning to address a range of interconnected issues.

The ambition to create a more nature-friendly environment, for example, is evident not only in the extension of woodland and meadow planting but also in the decision to cultivate perennial flowers. Where digging over the earth to "plant bulbs constantly and stuff like that" resulted in the "degradation in the soil", the shift to planting perennial flowers aimed to support soil ecology rich in nutrients for "insect life", which, in turn, nurtures birdlife.[59] The selection of plants that bore more seeds on their flowerheads also predicted "more birds…coming to eat them".[60] Moreover, the sequential consequences of planting decisions rippled out from "tiny" to the "human scale", as "natural birdsong means you can't hear the traffic", enhancing relaxation in gardens enclosed by a busy road.[61]

The interrelation between planting biodiversity and the sustainable city also encompasses the need to mitigate against heavy rainfall, for Emma Friedlander-Collins observed the "one-in-100-year events" when "you get huge amounts of water falling out of the sky" has become "one in every ten":

> So, you need to build facilities in that will catch that rainfall. So, kind of sumps and stuff, but the more planting you can have that can filtrate through and catch that rain then the better. So, the planting does two jobs. It helps mitigate that rainfall, but it also is self-seeding.[62]

The aspiration to improve biodiversity was, in part, inspired by the history of the gardens during World War Two when the green was "turned over to nut tree planting" to produce "extra protein".[63] Where the aim was to support citizens' nutrition, the gardens, "full of walnut trees and almond trees," also afforded "really diverse" wildlife.[64] The ambition to increase food self-sufficiency in cities has been raised in our other case studies and it is important to recognise the historical antecedents for such action, even in smaller and more centrally located common spaces such as Valley Gardens. Thinking about Valley Gardens as a space for urban cultivation led Emma Friedlander-Collins to speculate whether in the future there could be:

> More spaces for food growing, local food growing and consumption, which I think would be a really interesting thing to see happening in a space like this.[65]

Where there are no current opportunities for food cultivation in the gardens, there are community composting bins located on the Level. Through a partnership between the City Council and the charity, The Food Partnership, local residents, typically flat dwellers with no outdoor space, can deposit their food waste in composting bins. The resulting compost is used by members and on the parks and gardens.[66]

In addition to the environmental reasons, the inclusion of meadow grasses and wildflowers, and the flows of planting aspired to generate a "really accessible and tactile" environment that invoked "an invitation to actually come and interact with it".[67] The observation of a child meandering open-armed into the flow of flowers stood in contrast then to the fenced off flowerbeds of the earlier ornamental gardens.[68] Moreover, the mapping of desire lines, the routes people chose to travel along rather than the designated footpath in the gardens, informed the placement of new footpaths. Lined with planting and lights, "so people feel comfortable to walk through these spaces"; these form the "flows" of movement through the common space. Looking out onto the gardens from the window of the university where Emma Friedlander-Collins now works, she observed:

> It's really incredible watching how people are using it, and that's exactly what you see, you see people walking the whole length of it, whereas before it was so shut off.[69]

The invitation to walk across or along the length of the gardens links back to another priority of the redevelopment team in their overture for how Valley Gardens might support a sustainable city:

> We wanted to make Brighton a city that people wanted to walk in instead of drive in. It was a real sustainability drive. And that was where all of this work originally stemmed from. How do you get people to stop driving into Brighton and make it a navigable, walkable, cyclable city?"[70]

Through the physical removal of barriers and the way design features offer an invitation to the public to come in, encounter and ultimately appropriate, Valley Gardens has opened into a common space for walking through and lingering in.

The ambition to generate an open common space for the future also extended to "more of that kind of public political space",[71] which chimes with the history of urban commons as sites of protest. Speaking of the role Valley Gardens plays in enabling citizens to congregate and publicly voice concerns, Emma Friedlander-Collins recalled various recent gatherings:

> So, we were talking about the gathering at the far end with all the candles and the vigil, and I think during, only a couple of weeks ago at COP26 The Level was the space where everyone met to start the kind of parade. The RSPB actually set up a big sustainability walk that started here.[72]

Her observation illustrated the range of issues which draw citizens together to collectively demonstrate the climate crisis more broadly and wildlife protection more specifically, while the vigil reflected both a coming together to mourn the death of Sarah Everard and a statement on women's safety.[73] Urban commons, as we have seen across all our case study sites, have historically continued to facilitate public rights to protest. In the past, Valley Gardens, or more specifically The Level, has accommodated various forms of protest, the violent clash between anti-fascists and the Union Movement in 1943 and the over two-month occupation by the Brighton Women's Peace Camp in 1983,[74] a couple of twentieth-century examples. In the present, Emma Friedlander-Collins identified Brighton as somewhere which is "full of activists", people who are aware of the sorts of issues raised above, who need spaces for public meeting and address.

Another expression of support in Valley Gardens as a site for public protest came in response to one of our participatory research creative investigations.[75] Tasked with the exploration of public signage, hence regulation in the Gardens, and how users either adopt the rules or not, parent and child participants, Elisha and Yali, reflected on conflicting ethical principles behind fly posted Extinction Rebellion posters announcing a forthcoming demonstration. In their reflective poem, seen below, Elisha and Yali, seemed to indicate that where pasting posters over City Council notice boards contravened the "rules", the urgency of the "climate collapse" message justified this individual action and the action of protest itself.[76]

In images generated through their investigation, Elisha and Yali also revealed numerous types of signs – a heart carved in a tree, graffiti, fly posters and public information notices. Of note was the way city council signage focused users' attention on what was not legally permissible rather than what was. It was perhaps not surprising, then, that where users often commented on how demonstrations were located in, travelled to or departed from The Level, none understood it to be one

Figure 5.5 Elisha and Yali. *A Common of Rebels.* In-Common Sites co-inquiry No. 2, Valley Gardens, Brighton, 2021.

Image: Siobhan O'Neill.

of the "set aside sites" where public meetings could be held, according to Council bye-laws.[77] Where the right to meet in this common-like space may not be fully understood, however, the reiterative performance of public protest serves, Emma Friedlander-Collins suggested, to keep the invitation open:

> So, I also think that maybe as those things become more and more important for us, actually we're going to see more of it being used for that kind of gathering with a kind of political intent or a community intent. And I think those spaces start to identify movements like that a bit. I think they can sort of start to shape how things go forward … what I'm hoping is the invitation stays open for people to then come and have those sorts of important gatherings.[78]

A similar sense of urgency is evoked in Emma Friedlander-Collins's perception of the growing importance of current issues, while the invocation of hope both points towards an aspiration and alludes to the possibility of failure. Ambiguity over the future of protest in urban commons, and open spaces more widely, has been driven by the government's Police, Crime, Sentencing and Court Bill.[79] The Bill has been criticised, including at the so-called "Kill the Bill" demonstrations on The Level, for placing greater restrictions on people's right to protest. In this context, then the continued hope for urban common spaces to extend an invitation to political gathering and for people to be not afraid to occupy these spaces is compelling. For it was not simply any public space that Emma Friedlander-Collins perceived as significant in this role, it was green space, which she evocatively invoked in her description of a "massive tree" in the gardens as the "focal point" for vigil and ongoing memorial for Sarah Everard.[80]

> So, this tree has no idea that that's what it's being used for, but it's now this kind of, this anchor for people to come and gather around and share something through its physicality and its, and I wonder if it would be the same if it was a different object that was there. I think it probably wouldn't. I think a tree in and of itself feels quite an inviting object you want to interact with because we've got other things in the space that are manmade, and nobody stood around.[81]

Notes

1 Charles Webb and A. E. Wilson, eds., *Elizabethan Brighton: The Ancient Customs of Brighthelmston 1580.* (Brighton: John Beal and Son Ltd., 1952).
2 Ibid., p. 28.
3 Sue Berry, 'Pleasure Gardens in Georgian and Regency Seaside Resorts: Brighton, 1750–1840', *Garden History*, 28:2 (2000): 222–230.
4 John H. Farrant, 'Russell, Richard (1687–1759), Physician', *Oxford Dictionary of National Biography.* Available at: https://www.oxforddnb.com/view/10.1093/ref:odnb/9780198614128.001.0001/odnb-9780198614128-e-56302 (Accessed 15 June 2022).
5 John A. Erredge, *The Ancient and Modern History of Brighton*, with a reprint of "The Booke of all the Auncient Customes, 1580" (Brighton: W. J. Smith, 1867), p. 189.
6 Henry Martin, *The History of Brighton and Environs* (Brighton: John Beal, 1871), p. 44.

7 R. Sickelmore, *History of Brighton and its Environs*, from the earliest period to the present time (Brighton: C. & R. Sickelmore, 1827), p. 31; Martin, *The History of Brighton and Environs*, p. 6.
8 Ken Fines, *A History of Brighton and Hove. Stone-Age Whitehawk to Millennium City* (Chichester: Phillimore & Co. Ltd., 2002), p. 5.
9 *Domesday Book: A Complete Translation* (Harmondsworth: Penguin, 1992), pp. 63–64.
10 *The Ancient Customs of Brighthelmston*, 1580, ed. Webb and Wilson, p. 33.
11 Ibid.
12 Sickelmore, *History of Brighton and its Environs*, p. 32.
13 Ibid.
14 *Brighton Directory* as quoted in Royal Pavilion Brochure, p. 2.
15 Royal Pavilion Brochure, p. 44.
16 Erredge, *The Ancient and Modern History of Brighton*, p. 195.
17 Berry, 'Pleasure Gardens in Georgian and Regency Seaside Resorts', pp. 222–230.
18 Erredge, *The Ancient and Modern History of Brighton*, p. 186.
19 Morning Herald, 9 and 19 August 1805. Erredge, *The Ancient and Modern History of Brighton*, pp. 186–187 and 202.
20 Morning Herald, 11 September 1805, quoted in Erredge, *The Ancient and Modern History of Brighton*, p. 186.
21 https://www.georgeglazer.com/wpmain/product/a-cricket-match-between-sussex-and-kent-victorian-colored-lithograph/
22 Erredge, *The Ancient and Modern History of Brighton*, p. 202.
23 Ibid., p. 296.
24 https://britishlistedbuildings.co.uk/101380932-old-ship-assembly-rooms-regency-ward#.Y4Xnzy-l2Tc.
25 Berry, 'Pleasure Gardens', p. 224.
26 Ibid.
27 Sicklemore, *History of Brighton and its Environs*, pp. 62–63; https://sbpc.regencysociety.org/lucombes-library-in-part-of-the-steine-brighton/
28 See: https://www.brighton-hove.gov.uk/sites/default/files/2022-05/Valley%20Gardens%20Conservation%20Area%20Study.pdf (Accessed 25 January 2023).
29 See ibid. at 3.2.3.
30 For a full list of the problems identified by the 1995 study, see ibid. at 10.1ff.
31 See ibid. at para. 3.5.10.
32 Specifically, Hanover Street and Terrace and Richmond Terrace, see ibid. at 7.8 and 7.9
33 Ibid. at 7.4, 7.5.
34 Ibid. at 10.3.
35 See https://www.brighton-hove.gov.uk/sites/default/files/migrated/article/inline/downloads/transport/Social_Emotional_Benefits_of_Good_Street_Design_Brighton_Hove_City_Council_September_2011_649KB.pdf
36 See https://en.unesco.org/biosphere/eu-na/brighton-lewes-downs
37 See https://www.brighton-hove.gov.uk/sites/default/files/migrated/article/inline/valley-gardens-master-phasing-plan.pdf
38 For a full description of the landscape design principles and their implementation at the site, see further https://worldlandscapearchitect.com/valley-gardens-a-new-1-5km-linked-park-system-in-the-heart-of-brighton/#.Y9Ei8C-l1mA
39 Graham Perring, "Beyond Public Space: Design-Make Pavilion_2019," *Studio Perring: Architecture–Research–Studio*, 14 February 2019. Avaliable at: https://studioperring.co.uk/studio/beyond-public-space-wastes-and-strays-design-make-pavilion_s06-2019/.
40 See https://www.brighton-hove.gov.uk/sites/default/files/migrated/article/inline/valley-gardens-master-phasing-plan.pdf
41 Walking Conversation with Jim Mayor on 25 June 2021.
42 Ibid.
43 Ibid.

44 Ibid.
45 Walking Conversation with Jim Mayor on 25 June 2021.
46 Walking Conservation with Emma Friedlander-Collins on 28 November 2021.
47 Ibid.
48 Ibid.
49 Walking Conversation with Jim Mayor on 25 June 2021.
50 See: https://www.brighton-hove.gov.uk/sites/default/files/2022-05/Valley%20Gardens%20Conservation%20Area%20Study.pdf
51 Walking Conservation with Emma Friedlander-Collins on 28 November 2021.
52 Ibid.
53 Ibid.
54 Walking Conversation with Jim Mayor on 25 June 2021.
55 Walking Conservation with Emma Friedlander-Collins on 28 November 2021.
56 Ibid.
57 Walking Conversation with Jim Mayor on 25 June 2021.
58 Walking Conversation with Jim Mayor on 25 June 2021.
59 Walking Conservation with Emma Friedlander-Collins on 28 November 2021.
60 Ibid.
61 Ibid.
62 Ibid.
63 Ibid.
64 Ibid.
65 Ibid.
66 See Brighton & Hove Food Partnership *The Level Composing Community.* Available at https://bhfood.org.uk/directory/the-level-community-composting/
67 Walking Conservation with Emma Friedlander-Collins on 28 November 2021.
68 Observation of public users at Valley Gardens, Brighton, 17 August 2020.
69 Walking Conservation with Emma Friedlander-Collins on 28 November 2021.
70 Ibid.
71 Ibid.
72 Ibid.
73 For details of the vigil held for Sarah Everard at Valley Gardens, see BBC News. Sarah Everard: Sussex Police jeered as park vigil in Brighton halted (*BBC*, 2021). Available at: https://www.bbc.co.uk/news/uk-england-sussex-56392225

For details of the Sarah Everard case, see V. Dodd and H Siddique. 'Sarah Everard murder: Wayne Couzens given whole-life sentence (*The Guardian*, 2021). Available at: https://www.theguardian.com/uk-news/2021/sep/30/sarah-everard-murder-wayne-couzens-whole-life-sentence

74 For more information on the Peace Camp at the Level, see Sam Carroll 'Brighton Women's Peace Camp, 1983: Second Wave Feminism and the Women's Peace Movement', *University of Sussex Journal of Contemporary History* 8 (Winter), (2005).
75 *In-Common Sites* co-inquiry No. 2 Signs, 2021.
76 Elisha and Yali, Participant Comment, *In-Common Sites* co-inquiry No.2 Signs, 2021.
77 In response to a 2011 Freedom on Information request, Brighton and Hove City Council released the *Pleasure Ground, Open Spaces, ETC. County Brough of Brighton Byelaws,* (1954). Available at https://www.whatdotheyknow.com/request/parks_bylaw#incoming-185368
78 Walking Conservation with Emma Friedlander-Collins on 28 November 2021.
79 Police, Crime, Sentencing and Courts Act 2022. See Part 3 Public Order. Available at https://www.legislation.gov.uk/ukpga/2022/32/part/3/enacted
80 Walking Conservation with Emma Friedlander-Collins on 28 November 2021.
81 Ibid.

Part II

The Urban Common as a Contested Common Space

6 Land Use and the Culture of Urban Commons

Continuity and Change

Introduction

Today it is the continuity of urban commons that tends to be celebrated. They are frequently presented by those who manage, use and study them as spaces that have survived against the odds – as remnants of the past in the present. For example, as Elizabeth Halcrow stated in her account of Newcastle's Town Moor:

> At one time there were a great many commons in England, but they disappeared steadily as the enclosure movement spread and as towns expanded and swallowed up all available land within and adjacent to their boundaries. Newcastle, no less than other towns, grew in size and importance, yet the Moor was preserved – intact and open, free from the encroachments of the builder or the enclosing zeal of the landlord which transformed open land in to private demesne - a unique and priceless heritage which has contributed to the enjoyment and well-being of generations of citizens.[1]

Similarly, the General Secretary of the Open Spaces Society, Kate Ashbrook, describes commons as "a survival from the Middle Ages which the public now has the right to enjoy".[2]

There is some substance to this view. Most surviving urban commons have existed as open spaces since medieval times or even earlier.[3] Some, like Newcastle's Town Moor, continue to be the focus for agricultural activities (in this case, cattle grazing) that have been crucial to their existence and history over many centuries.

Yet, despite their longevity, urban commons are not unchanging spaces. The historic record demonstrates that they have been subject to substantial change provoked by decisions at both a local and a national level. It is the aim of this chapter to celebrate the continuity of these spaces, whilst acknowledging change as an inherent feature both of the landscapes themselves and of the ways in which they serve local communities. This chapter will first explore continuity and change in relation to the space and extent of urban commons through consideration of the impact of enclosure and other forms of encroachment. Attention will then turn to how the uses to which urban commons were put adapted over time. By understanding

DOI: 10.4324/9781003204558-8

the resilience of urban commons to historic change, we hope to open dialogue on the future of these spaces as sites of balanced adaptation and protection.

Enclosure and the Loss of Urban Commons

The urban commons that still exist today cover a tiny fraction of the area that once constituted our common land. Whereas in medieval times approximately half of England could be considered common land, today that figure is just 3%.[4] In large part, this is due to enclosure: the appropriation, fencing off and cultivation of land that was previously waste or common land with the associated loss of rights previously held by local commoners.

The origins of enclosure can be traced back to the Statute of Merton of 1235. It declared that enclosure of commons was allowed, so long as sufficient land was left for commoners. This statute was used by lords of the manor to justify enclosure for centuries thereafter.

The history of enclosure in England and Wales has been divided into three distinct periods.[5] The first extends from the Black Death in the mid-fourteenth century to the early eighteenth century. While this was the period in which the majority of English land was enclosed, it was primarily agrarian land that was enclosed for agricultural purposes. Urban commons were less susceptible to enclosure, and around the capital, commons were explicitly protected by the law of 1539 which prevented the enclosure of commons within three miles of London.

Urban commons were more likely to be enclosed during the second period, stretching from 1750 until 1845 and dominated by parliamentary enclosure, initially by private parliamentary bills. The historian Henry French, drawing on the survey lists of parliamentary enclosure acts published in 1914, demonstrated that between 1720 and 1870, at least 160 towns sought enclosure acts.[6] This is probably an underestimate, since various cases, including enclosures of London commons, were not included in these lists. These acts resulted in considerable loss of commons and wastes in response to "improving" agricultural practices and rapid urban expansion. This style of enclosure proved detrimental to urban and peri-urban commons, many of which were converted to urban uses despite being valued as spaces of continued rural practices. A good example of this enclosure is the case of Nuns Green in Derby, an area of land of approximately 50 acres over which local inhabitants had common rights. Local commoners had encroached on the space themselves by digging for gravel or erecting small buildings on the land. In 1768, an act was passed to sell a portion of the Green so that houses could be built on it. This would both address a need for housing and frustrate the actions of commoners. Just over 20 years later, in 1791, it was decided that the remaining portion of the Green would be sold for the same purpose with the proceeds of the sale being used to fund the paving and lighting of the town of Derby. The application to Parliament prompted opposition as reflected in cheap publications including the song "Quoth old Ned" which began:

> 'Quoth *old Ned*, to his *Lad;*- "I have been told,
> *Nun's-Green*, my little dear, is to be sold

To pave, and light, old Derby; (*fulsome town!*)
And save the POOR from *laying money down.*
Now is it *fair*, that you and I should be
Depriv'd of our just *Rights*, and *Property*?[7]

Despite this opposition, Nun's Green was sold.

By the 1830s and 1840s, concerns were being raised from various quarters about the (mis)use of urban commons. Many commons that had not been enclosed by this point were neglected, meaning they were often overgrown and/or subject to violations. At the same time, rapid urbanisation had brought large numbers of workers into towns and cities and there was a growing concern that these people needed accessible outdoor spaces for recreation and exercise. In a debate in the House of Commons on 21 February 1833, Robert Slaney proposed the establishment of a Select Committee "to consider the best means of securing open places in the neighbourhood of great towns for the healthful exercise of the population".[8] The Select Committee on Public Walks was duly established later that year to address the poor environmental conditions of England's urban areas, specifically by considering "the best means of securing open spaces in the immediate vicinity of populous towns, as public walks calculated to promote the health and comfort of the inhabitants".[9] The Committee acknowledged that much open space in towns had been enclosed in recent years and that "little or no provision has been made for public walks or open spaces, fitted to afford means of exercise or amusement to the middle or humbler classes".[10] Perhaps, not surprisingly, the concerns of MPs on this issue were not purely altruistic. The Committee believed that "some open places reserved for the amusement (under due regulation to preserve order) of the humbler classes, would assist to wean them from low and debasing pleasures" such as drinking, dog fights and boxing matches. The Committee was particularly concerned about manufacturing workers who were shut up in hot and dirty factories all week. If these workers were not provided with opportunities to enjoy their day of rest in the fresh air, the fear was that:

> their only escape from the narrow courts and alleys (in which so many of the humble classes reside) will be those drinking-shops, where, in short-lived excitement, they may forget their toil, but where they waste the means of their families, and too often destroy their health.[11]

Interestingly, the commentator in the *Westminster Review*, while largely positive about the Committee's report, did note that what was required was not merely gravel walks, but "public grounds" which would afford the opportunity not just to promenade, but for more vigorous activities such as football and cricket.[12]

These debates resulted in the 1836 General Enclosure Act, which included a public interest clause preventing the enclosure of open fields within a ten-mile radius of London. Yet this Act was not viewed positively by everyone. *The Spectator* argued that despite its intentions, the bill actually threatened the enclosure of Hampstead and Putney Heaths, Blackheath, Wimbledon, Roehampton, Wandsworth and

Clapham Commons, as well as Clifton and Durdham Downs near Bristol.[13] Enclosure of wasteland threatened "the lungs of great cities", such as London, Manchester, Sheffield and Bristol. The nub of the issue was that enclosure could not go ahead without the consent of the lord of the manor. Yet, as the author of the article noted, it was these landowners who had generally been the loudest advocates for enclosure in the past.[14] There were also growing concerns that the process of securing enclosure through private parliamentary bills was complex, time-consuming and often antagonistic.

To address these issues, the 1845 Act created a permanent "Inclosure Commission" that did away with the need for individual acts and set out a clear process with a body of assistant commissioners who could support those seeking to enclose an area of land. This Act also formalised the provision of compensatory land when common land was enclosed. This did go some way towards limiting enclosure, but as Alun Howkins has shown, enclosure did continue post-1845. He estimated that there remained approximately 2.5 million acres of common land in 1845, which had been reduced to just under 1.9 million acres by 1914.[15] Included within this was approximately 4,100 acres of common land in London that were regulated under the Metropolitan Commons Act between 1866 and 1898. Howkins also noted a shift during this period, with strong support for enclosure in the 1840s giving way to a slowdown in the 1870s as a result of political opposition, as well as a shift after 1876 towards regulation rather than enclosure of the commons. The period after 1845 also saw more land being enclosed for urban development rather than agricultural improvement. Urban authorities were particularly keen to develop land that was viewed as unproductive. Encroachment at the edges of green spaces became commonplace, as housing or railway lines became legitimate uses of such land.

These changes after 1845 coincided with, as well as sometimes being driven by, a more concerted campaign against enclosure, epitomised by the establishment of the Commons Preservation Society.[16] In 1864, Earl Spencer put forward a bill proposing that part of Wimbledon Common be enclosed. This led Frederick Doulton MP to establish a Select Committee: "To inquire into the best means of preserving for the use of the public the Forests, Commons and Open Spaces in the neighbourhood of London". The Committee proposed that the law be amended to prevent further metropolitan enclosures, resulting in the Metropolitan Commons Act of 1866, which dictated that commons within the Metropolitan area be managed by conservators elected by ratepayers. Yet this proposal also sparked action by various lords of the manor who began enclosing commons, fearing that they would not be able to do so for much longer.

It was these actions that led George Shaw Lefevre to hold the inaugural meeting of the Commons Preservation Society at his chambers in the Inner Temple on 19 July 1865. The aim of this society was to defend open spaces, especially commons, in urban areas by "the rousing of local opposition to proposed enclosures, the raising of funds for legal battles against landowners, and the collection of the legal and historical material necessary to evince the case for new legislation to protect common lands".[17]

The Society boasted a number of high profile members, who skilfully deployed their connections with lawyers and local landowners to great effect. Yet, at the same time, they did not shy away from powerful symbolic action. One of the Society's first campaigns concerned encroachments on Berkhamsted Common, where 434 acres had been enclosed within five-foot high iron fences by the trustees of the second Earl Brownlow, lord of the manor and owner of the neighbouring Ashridge estate. A local churchwarden William Langman wrote to *The Times* about the enclosure and then approached the CPS. Shaw Lefevre persuaded his friend Augustus Smith, who owned land nearby, to get involved and they co-ordinated a deliberate act of fence-breaking, arranging for more than 100 navvies to travel from London in the middle of the night for the purpose. After a protracted legal battle, the case was resolved in Smith's favour in January 1870.

The CPS was also involved in the battle to save Hampstead Heath in 1868 and in campaigns relating to Wimbledon, Wandsworth and Plumstead Commons and Epping and Ashdown Forests. Ben Cowell argues that the model deployed in Berkhamsted was repeated elsewhere. The Society brought the enclosure to public attention, persuaded a powerful local resident to bring a legal challenge against the lord of the manor and secured expert legal advice to maximise the chance of success. They also engaged in extensive legal and historical research, including gathering oral testimony from commoners and examining earlier disputes in detail. Soon the CPS gained statutory representation on the Parliamentary standing committee on enclosures. While the contributions of the CPS were undoubtedly important in saving common land from enclosure, local people, including members of the working class, also played their part, as Mark Gorman has shown for Epping Forest.[18] Moreover, as these examples suggest, the actions of the Society were largely concentrated in the south. This reflects the fact that campaigns relating to urban commons in and around London tended to receive greater national attention than those beyond. At the same time, urban commons located further from the metropolis may have been less subject to pressure for development (though northern industrial cities are perhaps an exception in this regard).

Important actions followed. The 1876 Commons Act (passed in the aftermath of the Epping Forest case) asserted that the wider "public" benefits of common lands be considered in all enclosure cases. Then, in 1893, the Statute of Merton was finally repealed. And in 1925, the Law of Property Act declared the legal right of access to all commons and waters in public areas. This included the right of the public to walk and ride on commons for the purposes of "air and exercise" (section 193) and the protection of many commons from encroachment and development (section 194).

The focus of the rest of this chapter is on those urban commons that resisted enclosure and survived. Even where this was the case, the shape of these commons did not remain static. Rather, they proved susceptible to encroachment of various kinds. Precisely how encroachment impacted on the space varied from one common to another, depending on a number of factors. In the first place, encroachment that was initiated or endorsed by the authorities was much more likely to be successful and durable than that advocated by commoners. Landowners and local authorities were also better placed to reverse or adapt encroachments that came from below. One example of this is the 1768 Act to sell a portion of Nun's Green in Derby mentioned above, which was a deliberate move to frustrate the activity of commoners who had erected a number of small buildings on the land.[19] Encroachment could also vary simply on account of local circumstances and concerns. We can, however, identify three broad types of encroachment: the permanent loss of land around the edges of the common which changed its extent and shape; the erection of permanent (or semi-permanent) buildings or other structures on the common; and the construction of roads across the common. Again, each of these methods operated differently – and with strikingly varied results – in each case.

Continuity and Change in the Shape of Commons

The Newcastle Freemen claim that since the Town Moor Act of 1774, just 210 acres have been eroded from the Town Moor leaving just under 1,000 acres of the original space intact.[20] Much of the encroachment that has occurred has been focused on Castle Leazes where since the early nineteenth century a barracks, a park, a hospital and university buildings have been constructed. Development on the Leazes no doubt helped to protect other parts of the Town Moor from encroachment, but another factor in the protection of the space has been the tenacity of the Freemen who have fiercely contested each and every attempt to diminish the Moor.

In Bristol and Brighton encroachment has arisen primarily due to increased urbanisation and development during the eighteenth and nineteenth centuries. In both cases, this was due to the rise of the surrounding area as a spa resort. Yet the precise impact on the shape of each common has been different. Clifton and Durdham Downs were originally separate from the city of Bristol, with agricultural land in between. The development of the suburb of Clifton as a fashionable residential area associated with the rise of Hotwells Spa, however, brought the city to the edge of the Downs, turning it into a recreation area for local residents. Yet, while the city expanded up to the Downs, there was relatively little encroachment onto the common itself, with the extent of the Downs remaining fairly constant over the centuries.

In Brighton, the Wellesbourne Valley originally marked the edge of the urban centre, as is evident from the image of a naval attack on Brighton (Chapter 5, Figure 5.2). Yet Brighton's reinvention as a resort brought extensive development of housing to the east of the valley (see 5.3 in Chapter 5). As a result, the Valley Gardens area was transformed from a common on the edge of the town to one at its heart (though the new houses themselves did not encroach on the common) and as with Clifton and Durdham Downs, it became a crucial recreational space for residents and visitors alike.

The encroachment of Mousehold Heath was both more extensive and of a different nature. The Mousehold Heath of today originated as a small corner of a vast area of common land that stretched for 6,000 acres from the medieval city of Norwich out to the Norfolk Broads. It was not then originally an urban common as such, but has become so due to encroachment, since it is the portion closest to the city that remains. Moreover, for the most part, the reduction in space was due not to the expansion of urban settlements, but rather to agricultural enclosure affecting the rural parts of this once extensive Norfolk common. John Wagstaffe's 1792 proposal marked the beginning of more extensive encroachment on this space. Wagstaffe placed emphasis both on the greater productivity of the land that would arise from enclosure and cultivation, but also suggested that the part of the Heath within the Norwich boundary be converted into pleasure grounds for exercise and recreation.[21] The enclosures that followed as a result of the early nineteenth-century acts were swift and brutal with only around 190 acres remaining at the Norwich end by 1838. The encroachments that occurred in the late nineteenth and twentieth centuries – which are detailed in Figure 3.3 in Chapter 4 – were, by contrast with earlier examples, primarily for urban uses, including for new housing and for military and carceral establishments (which are discussed in more detail below).

While the exterior boundaries of these urban commons may have remained relatively constant over the centuries, the organisation, structure and appearance of the space within has been subject to significant change.

Buildings and Structures

It is sometimes difficult to distinguish between encroachment that nibbles at the edges of a common and that associated with the erection of permanent (or semi-permanent) buildings and structures, since at least some buildings were constructed at the margins of the land. Sometimes, these encroachments were made by commoners themselves. Nun's Green in Derby was used during the 1660s as a space for temporary market stalls during the plague. Farmers were unwilling to enter the town due to the risk of infection, and this provided a way to continue the trade and maintain trust between buyers and sellers. Another example is the shelters that were erected on Warminster Common in Wiltshire in the mid-nineteenth century, which provided accommodation for poor locals as a kind of proto shanty town. When the residents refused to give up their dwellings, they were accused of illegal occupancy by the local authorities. As in the case of the quarrying undertaken on Mousehold Heath by the residents of Pockthorpe the inhabitants of the Warminster shelters were characterised as immoral, un-Christian and inclined towards criminality.

Generally more successful and more enduring were structures erected by the authorities themselves. The barracks erected on part of Castle Leazes in Newcastle in the early nineteenth century and the hospital situated there at that century's end are prime examples of this. Indeed, the Leazes was the most vulnerable part of the Town Moor, precisely because it abutted the built environment of the city. Yet in each of our case study locations, there were also buildings erected more centrally on the commons either to benefit from the isolated nature of the spot or to facilitate recreational or other activity in the space.

The location and open nature of urban commons – separated from inhabited areas by open land, but not too far distant – made them an ideal place to build isolation hospitals. The leper hospital of St Mary Magdalen was established on

Mousehold Heath in 1119 and survived there (later operating as an almshouse for the poor, sick and elderly) until the sixteenth or seventeenth century.[22] The site of the isolation and smallpox hospital on Newcastle's Town Moor is now marked by a fenced enclosure containing a small copse of mature trees and scrub close to the two artificial hills that resulted from the construction of the urban motorway in the 1970s.[23] The hospital opened in 1882 and was divided into two separate units that were set within a walled enclosure. One was a smallpox unit with beds for 72 patients, the other a more general isolation ward with 100 beds. The Freemen only agreed to the building of the hospital on condition that they were compensated by the incorporation of a similar sized piece of land into Nuns Moor close to Fouracres Road, thereby ensuring that the building did not result in a reduction of the overall size of the Moor. The hospital buildings were permanent, constructed out of timber and corrugated iron. Images from the late nineteenth and early twentieth centuries provide a sense of both the setting of the enclosure within the landscape of the Moor and of the internal layout of the wards (see https://co-curate.ncl.ac.uk/smallpox-hospital-town-moor/).

The hospital was demolished in the mid-twentieth century, with the last buildings being removed in the 1960s not long before the creation of the urban motorway. Soon after the demolition, the eastern end of the enclosure was extended and the trees planted. Two paths stretching from Grandstand Road to the edge of the enclosure survive.

The hospital that opened in 1882 was almost certainly on the site of an earlier isolation hospital built at the time of the cholera epidemic of 1866. The Sanitary Act 1866 gave town councils and local boards of health the power to build temporary or permanent hospitals, with Justices of the Peace being empowered to move patients to them.[24] In the end, only a small number of hospitals were constructed before the epidemic subsided, but Newcastle's Town Moor does appear to have been the location for one of these. The *Newcastle Journal* for 1 September 1866 noted the erection of the hospital by the Public Health Committee "mid the desolate waste of the Town Moor" to be used in case an epidemic were to break out in the city.[25] It was said to be located a quarter of a mile to the south-west of the Grandstand and consisted of a day room, nurses' room, kitchen, offices and a ward to hold 80–90 patients. It was isolated from all public footpaths and buildings and had good drainage.

Hospitals were not the only kind of municipal facility that suited the isolated location of urban commons. This could be a benefit for prisons too, with urban commons offering a secure location in which inmates could be easily contained, and often conveniently located close to sites used historically for judicial corporal punishment. Nun's Green in Derby had long been used for executions and this no doubt contributed to the decision to build a prison there in 1756. Moreover, there is also evidence of an earlier prison on the site.[26] Norwich prison was opened in 1887 on a southern section of Mousehold Heath. Since the fourteenth century, the county gaol had been located at Norwich Castle, but as early as the 1820s, this facility was no longer deemed adequate. Mousehold Heath offered the perfect alternative site – large enough to house a modern facility and provide adequate security (the building was set within substantial grounds surrounded by high walls), but still close enough to the city centre to make the transportation of prisoners relatively easy. The new purpose-built complex was opened on 2 August 1887 and subsequently extended in the 1960s, taking over more of the Heath.

During and after World War Two, structures were also erected on urban commons to house prisoners of war (PoWs). A PoW camp was erected at the eastern end of Nuns Moor in Newcastle, to the north of Studley Terrace, to house Italian prisoners.[27] It consisted of a series of prefabricated buildings comprising several parallel accommodation blocks and other facilities. As with Norfolk prison, the location made it possible to enclose the building within a perimeter fence and to provide further defences in the form of several trenches down the west side (which appear on aerial photographs from 1947 and are still visible in the landscape today). Despite these defences, the prisoners were not permanently confined to the camp, but were allowed out to carry out jobs around the town, including road sweeping and delivering mail. The camp was demolished in 1959 and returned to grassland.[28]

A PoW camp was also established on Mousehold Heath.[29] The area in which it was located, to the east of a quarry, had been used for a variety of military purposes between 1917 and 1951, including others that involved the erection of temporary structures, such as the tented camp in place from 1917, which comprised five rows of tents and four of Nissen huts. The PoW camp was in place by February 1946 and comprised at least 21 Nissen huts of varying lengths and several ancillary buildings. Tents and at least 24 more Nissen huts were added soon after to provide additional accommodation. It is not certain whether this was the camp listed as no. 253 in the official list of PoW camps, which is described as a German working camp, but is said to have been located approximately 700 metres to the north-west of this site. As in the Newcastle case, there was clearly some integration with the local community. The Norfolk Record Office holds an interesting collection of letters belonging to Mrs Kathleen Statham of Norwich.[30] The first, dated 18 December 1947, is from the commandant of the camp on Mousehold Heath giving permission for a prisoner, Hans Dittrich, to visit Mrs Statham on Christmas Day 1947. A second letter, dated 20 December, is from Dittrich accepting the invitation. The Norwich International Fellowship Committee was responsible for arranging for prisoners to spend Christmas day with local residents, but Dittrich and Statham clearly hit it off and continued to communicate until at least the end of 1950 after Dittrich had returned to Germany. After the Christmas visit in 1947, Dittrich wrote to Statham on 7 January 1948 with an invitation of his own:

> There is a concert of a P.o.W. – String-Quartet at our camp on next Saturday (January 10) at 7 p.m. Our Commandant has given permission to invite 100 guests to that performance. May I invite you, Muriel and Audrey, if you haven't any plans for Saturday yet? The program includes works of Bach, Schubert, Mozart, and Haydn and will last about one hour and a half.
>
> I shall meet you at 6.45 at the 92 – Bus stop at Harvey Lane, if you will come. I should be really pleased to see you on Saturday.[31]

Alongside PoW camps, other structures to aid the war effort were constructed on urban commons during the two world wars. The large open space made them particularly valuable for the building, maintenance and storage of large military vehicles. An aircraft factory was established on Newcastle's Town Moor and a tank repair depot and military vehicle park on Bristol Downs. The aircraft factory took over the racecourse grandstand on Town Moor. The practice of commandeering existing buildings for the war effort also occurred in Bristol, where changing rooms on the Downs were repurposed as a gas detection cleansing station and base for training in decontamination. Urban commons were also ideal locations for the establishment of structures associated with defence and the Town Moor, Mousehold Heath and the Downs were all used in this way.

The building of military structures on urban commons was not an innovation of the twentieth century, but had a much longer history. Arsenals were located close to Valley Gardens in Brighton at various points from the sixteenth century and temporary militia camps were erected on the Steine in 1784 and 1796 (as seen in Figure 6.1).

Both Town Moor and Mousehold Heath became locations for more permanent military facilities in the nineteenth century. In 1805, the Mayor and Freemen of Newcastle granted five acres of the Leazes to the government at an annual rent of £5 per acre for an artillery depot as part of the defensive campaign associated with the Napoleonic Wars. The site soon developed into a barracks, with two stable blocks being constructed that could accommodate more than 180 horses. Barrack rooms and fine houses for officers were built along with cooking and cleaning houses, forage sheds and offices. Later, a canteen and an officers' mess room were added. Hospital facilities were also provided for both men and horses. The buildings ended up costing more than £40,000.[32] In Norwich, two sections along the southern boundary of Mousehold Heath were obtained by the War Office in 1884

Figure 6.1 Old Steine, Brighton, from the North, 1796.

for the construction of what became the Britannia Barracks, the headquarters of the Royal Anglian Regiment. Owing to their proximity to the town, it was requested that the barracks be aesthetically pleasing and the result is said to be one of the most architecturally sophisticated nineteenth-century barracks in the country.[33]

While the War Office could overturn rules preventing the erection of permanent structures on urban commons, these rules did obstruct the construction of other sorts of facilities, including those designed to service legitimate uses of the space by the public. In 1903, the Bristol Rugby Combination requested the construction of changing rooms on the Downs for the use of their members who played there. Since the erection of permanent structures on the Downs was forbidden under the 1861 Act, the Downs Committee could not immediately address the need. Requests for changing facilities continued to be made, but it was almost 30 years before the issue was finally resolved. Indeed, resolution only came about because the Bristol Water Works Company agreed to provide land adjacent to their reservoir – which had been purchased before the 1861 Act – for the changing rooms. The plan was further supported by a grant from the Parks and Open Spaces Relief Scheme made in 1930.

The strict prohibition on building in Bristol probably prevented the construction there of the sort of buildings seen on other urban commons, such as hospitals and prisons. Yet, Dorset House, located on an adjacent road, did make deliberate use of the open space of the Downs. Dorset House and its neighbour Alva House were a pair of villas originally built as part of a wider development by the Merchant Venturers to provide fashionable houses for wealthy merchants. They were adapted in 1929 to provide a treatment centre for patients suffering mental health problems and the first School of Occupational Therapy in the UK.[34] A residential clinic for women suffering mental disorders was located there from 1929 with the School following in 1930. Both were established by Dr Elizabeth Casson OBE (1881–1954) who had previously worked for Octavia Hill at another early occupational therapy centre, Red Cross Hall.[35] In this case, the relevance of the location was less to do with isolating patients and more about being able to make the most of the open space of the Downs. Central to Casson's method was the importance of access to outdoor space, with countryside excursions central to the therapies offered alongside artistic practices, including dance and drama. The hospital flourished, accommodating 800 patients between 1929 and 1941. Both facilities operated until World War Two when they were temporarily relocated. Bomb damage led to the permanent relocation of the School in 1946. The restrictions on building on Clifton Down resurfaced again in March 2023, when it was reported that the 1861 Act was hindering the redevelopment of a toilet block on Clifton Down to include a café alongside refurbished toilet facilities. It was reported that the Council would require special permission from Parliament to override the terms of the Act.

Perhaps the most significant building to encroach upon the space of Valley Gardens in Brighton was of a rather different kind. St Peter's Church was built in 1824 as part of the improvement scheme for the area and to address the acute shortage of churches in Brighton resulting from the rapid expansion of the urban area and consequently of the population. The fact that the expansion of the town had largely been to the east and the north presumably made Valley Gardens an ideal location. A competition was launched in August 1823 to determine the design of the church, which was won by Charles Barry (a rising British architect, who later penned the design for the rebuilt Houses of Parliament following the fire that destroyed the original building in October 1834).[36] The church was constructed of Portland stone in the Perpendicular Gothic style with a large tower that was deliberately designed as an impressive landmark at the top of the Steine.

Roads and Pathways

It is not only buildings but roads too that have encroached upon urban commons. In general, this occurs for one of two reasons. Either the roads are needed to allow communication across the space or they are a means of facilitating recreational activity within it. A number of examples of the former can be identified among our case studies.

In the late seventeenth and eighteenth centuries, action was taken to create a network of well-maintained roads throughout England in order to improve the transportation of people and goods across the country. Turnpike Acts gave local trusts the power to levy tolls on those using a particular stretch of road (usually about 20 miles) with the income ploughed back into repairs and improvement. The "turnpike" itself was the gate through which travellers passed having paid their due. The first Turnpike Act was passed by Parliament in 1663 and applied to a section of the Great North Road between Hertfordshire and Huntingdonshire.[37] Though the take up was initially slow, by the eighteenth century several roads were being turnpiked each year.

Turnpikes were constructed on several urban commons (often following traditional routes) as a means of improving communication into and out of the city. Bristol Corporation began investing in them from the 1720s in a bid to make Bristol a local communication hub. As part of this investment, tollgates were erected on Stoke Road and Westbury Road, which ran across the Downs. This initiative proved controversial, and by the 1740s, repeated damage to the tollgates meant that the income from tolls was insufficient to keep up with the repairs, meaning that the roads were not properly maintained. The turnpikes constructed across Newcastle's Town Moor faired a little better. In 1747, work began on the Morpeth turnpike, which followed the old cow causeway from Barras Bridge, on the edge of Newcastle city centre, north across the Town Moor and then on for approximately 20 miles to the Northumbrian market town of Morpeth. It has been suggested that the portion of the road across the Moor itself (which was subsidised by the Freemen) was better maintained than the section north of the Moor that was funded purely by tolls. Six years later, the Ponteland Turnpike was established. It ran west from Gallowgate via West Cowgate and on to the village of Ponteland just under ten miles north-west of the city.

Private individuals also attempted to construct roads across both Newcastle's Town Moor and Bristol Downs for more personal motives. In the 1760s, William Ord, brother of the former Mayor of Newcastle John Ord, applied to build a road from his residence at Fenham Hall on the edge of the Moor to the junction with the newly constructed Ponteland Turnpike. The Freemen accepted Ord's right to build a bridleway, but refused him permission to break the soil in order to construct a carriageway.[38] Newcastle's Common Council overruled the Freemen and allowed Ord to build his road and to hang gates in a hedge between his estate and the Town Moor to prevent access from the road onto his estate. While this decision stood, the construction of roads across the Moor continued to be of some concern, and in May 1771, a committee was appointed to examine the rights by which roads over the Moor were held or enjoyed. The Freemen were also concerned about other kinds of transport routes. In 1837, they successfully resisted plans for a railway line to be constructed across the Moor for the purpose of transporting coal. The proposal was rejected by "cow keepers" on the grounds that it would destroy the herbage.[39]

The attempt to build a private road across Bristol Downs in the nineteenth century proved equally controversial. In 1862, William Baker attempted to construct a road across Bristol Downs between what is now Ivywell Road and Worrall Road for the benefit of the residents of the new Sneyd Park area of the city. Though the Downs Committee initially approved the request and agreed to contribute to the cost, this decision generated a public outcry. A letter submitted to the *Western Daily Press* in 1868 was very critical of the management of the Downs over this issue:

> Where are the Downs Committee that they do not look after these things? Where is their man who is placed to look after the Downs to see they are not destroyed? And where are the police? Are they all blindfolded when they visit the Downs, are their visits nocturnal that they cannot see these things, or are they participators in this wholesale destruction of these our once delightful Downs, that were at one time the charm of the city and surrounding neighbourhood, but which are now turned into tramways, mud roads, and the grass everywhere being lost to view.[40]

While the Downs Committee paid to have Baker's road re-turfed, the route survived in the form of a footpath. The issue, however, remained deeply controversial. In 1875, the Committee explained that they could not build new roads due to a lack of funds. Two local men, E. S. Robinson and George William Edwards, offered to contribute £500 on behalf of the people of Stoke Bishop but only if Baker's road was reopened.[41]

Road building also impacted on our other two case studies. Gurney Road in Norwich, which cut across Mousehold Heath, was constructed in the 1880s as part of the wider improvement plans following the conclusion of the court case relating to the Heath. The Mayor noted that this "and all future work on the Heath" was to be aimed at "the preservation of the natural wild beauty of appearance".[42] In this case, the road appears to have been relatively uncontroversial, though the Conservators and Rangers did have to deal repeatedly with individuals damaging the banks of the new road as it was being constructed and with a local landowner taking his horse and cart across other parts of the Heath where there was no road.[43] The work of constructing the road became a public works project, with the town clerk H. B. Miller appealing to the unemployed men in the area to apply for the work with a promise that all tools would be provided and that payment would be 3d per hour for a six-hour day.[44] On 25 February 1886, it was reported that 195 men had been employed in preparing the ground. At the Mayor's request, an official opening for the road was staged in late May 1886.[45]

Brighton's Valley Gardens is now criss-crossed by roads making it feel more like a communications hub than an urban common. The origins of this layout can be traced back to the early nineteenth century, when the Prince Regent was granted permission to enclose London Road where it ran behind his Pavillion. Further development of the town to the east of the valley generated a need for more transport links, with various informal paths turned into new streets cutting across the space during the first half of the nineteenth century.

Perhaps the most substantial encroachment of a road across an urban common among our case studies is the construction of part of the urban motorway (now the A167) along the edge of the Town Moor and across Nuns Moor. Constructed in the early 1970s, this section of road was originally intended to form part of a larger motorway network, but the policy was changed before the full plan could be implemented. It was not just the motorway itself that transformed the appearance of the Moor, but also the creation of two large man-made hills from the spoil, one of which now marks the highest point on the Moor. Now popular with runners and for winter sledging, these hills are often assumed by visitors to be part of the natural topography of the space.

There are also several examples of roads being constructed across urban commons to facilitate recreational activity in these spaces. Newcastle's Grandstand Road, now a major artery bisecting the Moor and separating Town Moor to the south from Nuns Moor and Dukes Moor to the north, was built for the benefit of racegoers, providing access for carriages from the Great North Road at Blue House roundabout to the Grandstand itself. On Bristol Downs, both the Bridge Valley Road (constructed in 1822) and the two-mile long Circular Road (constructed during the 1870s) were designed to allow visitors to take in the Downs from carriages.

The Changing Uses of Urban Commons

It was not only encroachment that changed the character and appearance of urban commons. These spaces were also transformed on account of changing land use over time. The origins of urban commons were closely tied to the exercise of rights by commoners. Though grazing was the dominant common right exercised on most urban commons, there were also others, all of which impacted the landscape in particular ways. From the eighteenth century, these traditional uses were increasingly in competition with new uses of the space – both industrial and recreational – which brought changes of their own.

The character of commons as grass or heathland is directly linked to the grazing that takes place upon them, which prevents the regrowth of self-sown seedlings. The continued presence of cows on Newcastle's Town Moor has maintained its status as grassland over centuries. Sheep grazing was once common on Bristol Downs, and in the 1850s, the corporation deliberately made use of ancient grazing rights to protect Durdham Down from development.[46] Yet by the late nineteenth century, grazing had all but ceased on Clifton Down and only 300–400 sheep were being put out to graze on Durdham Down, with grazing ceasing there too by 1925. As a result, the grass had to be mowed to prevent regrowth and chemical fertilisers had to be brought in to replace the impact of sheep droppings. Since 2011, there has been a return to some grazing on the Downs, with goats being put out to graze the area of the gully to remove the brambles and thereby protect native plant species. On Mousehold Heath, where grazing no longer occurs, the ranger works with a team of volunteers to prevent the regrowth of trees in certain areas in order to protect the heathland ecosystem.

Despite the current perception of Newcastle's Town Moor as a grassland monoculture, there is a history of a changing landscape there too. Cultivation, native flora and mining all impacted on the appearance of the Moor over the centuries. Following the passing of the 1774 Act, portions of the Moor were temporarily leased out for cultivation. Though the precise regulations laid down in the Act do not appear to have been followed religiously – particularly the rule that no plot could be leased for a second time until the whole area of the Moor (with the exception of Cow Hill and the Race Course) had been let – the leasing out of plots did generate more than £1,300 for poor Freemen and widows between 1774 and 1812, and annual meetings to allocate plots were still being held into the 1870s.[47] These intakes must have created diversity in the use and appearance of the Moor even if grazing remained the dominant land use.

As well as grazing rights, commoners were also often entitled to estovers, which included the right to collect wood and furze for fuel. In the nineteenth century, certain areas of Town Moor were covered with whins or gorse, which was cut by some local residents to use for firewood. There might also be a right to take sand, gravel, stones or minerals from the common, though this was often vested in the owners of the land rather than the commoners themselves. There is evidence on surviving urban commons of sand and gravel pits dating back as far as the late medieval period, and coal was extracted from Newcastle's Town Moor almost continuously from then up to the twentieth century.[48] From the late seventeenth century, the lords of the manor of Henbury leased land on Bristol Downs for lead mining as well as for the quarrying of limestone and clay extraction.[49] Other minerals – including iron, manganese and calamine – were also extracted from the Downs, along with building materials for housing and roads. Building materials and resources for brickmaking were quarried on Mousehold Heath from at least the sixteenth century and the Stone Pit Company continued the practice into the twentieth century. Mining and quarrying inevitably contributed to constant changes to the landscape. On Bristol Downs in the late nineteenth century, quarries in certain areas were being opened at the same time as others were being filled. And in both Bristol and Norwich, quarrying produced direct material connection between the local urban common and the built environment of the surrounding area.

It was not only the exploitation of the finite resources of urban commons that impacted on their appearance. Urban commons were also ideal spaces for the generation, exploitation and storage of renewable resources such as wind and water, and while at less risk of overuse, the harvesting of these resources did affect the landscape. A windmill was erected at Clifton Down hillfort by James Waters in

1766–1767 at the behest – and with the financial backing – of the Society of Merchant Venturers. The mill was used to grind tobacco into snuff, with snuff-taking being a popular habit among both rich and poor in the eighteenth century. The windmill was destroyed by fire just ten years after its construction and was not replaced, but a ruin remained. Windmills are also evident in some of the images of the Old Steine in Brighton. Although these mills appear to have been situated up on the Downs rather than in the Wellesbourne Valley, they impacted on the appearance of the space, not least in enhancing its rural character.[50]

Drainage was often a problem on urban commons, not least in Newcastle and Brighton. In Brighton, the construction of a sewer in 1793, at the expense of the Prince Regent and Duke of Marlborough, was designed to resolve the problem. In Newcastle, attempts were made to turn a problem into an asset through the construction of reservoirs. In late 1767, Newcastle's Common Council established a committee to explore how the town might be better provided with water.[51] Following this, on 19 July 1770, the decision was taken to lease out a plot of land at the south end of the Town Moor for the construction of a new reservoir and to make 100 fireplugs that could be used to extinguish small fires around the city.[52] Some guilds remained concerned over such incursions. On 2 October 1775, the Cordwainers advanced money for the prosecution or defence of action at law relating to various issues associated with the Moor, including reservoirs and latrines.[53] In the early 1780s, a new waterworks was built and further reservoirs planned. Again the guilds expressed concerns about the damage to the Moor arising from the laying of pipes, insisting that compensation be paid to the Freemen for any damage to the herbage. These reservoirs were later transformed into lakes in Exhibition Park, marking a very direct repurposing from an industrial to a recreational function. A reservoir was also built on Bristol Downs during the 1840s following a good deal of debate over the supply of water to the expanding city, and in 1897, an agreement was made with commoners to drain part of Durdham Down.[54]

Though common rights could traditionally only be exercised for personal and domestic uses, not for commercial exploitation (as the Pockthorpe campaigners in Norwich found to their cost),[55] the boundary between the two could be unclear. While early coal extraction from Newcastle's Town Moor involved small quantities for purely domestic purposes, by the eighteenth and nineteenth centuries, mines were being operated by the Corporation. Yet, it was still argued that this was simply to supply the citizens of Newcastle with fuel.[56] On this basis, the exploitation of resources by public utility companies could also be justified.

These industrial uses of urban commons, which had exploited and even disrupted traditional common rights, were themselves displaced from the mid-eighteenth century by the reinvention of urban commons as spaces for recreation. Of course, it would be wrong to suggest that these spaces had not been used for recreational activities prior to the mid-eighteenth century. For example, in Newcastle an annual Lammas Fair was said to have been established by King John and a Cow Hill Fair occurred regularly from a later date. While these fairs were partly agricultural in focus, recreational activity was also involved. Yet from the mid-eighteenth century, a more formalised approach to recreation emerged, which was distinct from earlier usage.

In part, this was a direct result of increased urbanisation as new neighbourhoods and community groups sought to make the most of the green spaces in their vicinity and to call for the beautification of such spaces. In the case of Brighton and Bristol, this was closely tied to the establishment of spa resorts which brought residents and visitors alike to the area explicitly for the purpose of recreation. This transformation continued into the nineteenth century, encouraged by the concerns about the health and well-being of urban workers in the context of rapid urbanisation.

The rise of the notion that recreation was the primary function of urban commons brought with it a new emphasis on the aesthetics of these spaces. In Bristol and Brighton, the establishment of spa resorts prompted the development of new planting and landscaping schemes which, especially in Brighton, significantly altered the feel of the space. The improvement plans for Valley Gardens that were implemented in the 1810s and 1820s resulted in much greater landscaping and the division of the space into distinct portions, which (as images reveal) changed its character considerably. In Bristol, the new planting scheme of the 1870s, which included lining Ladies Mile with Huntingdon Elms, also impacted on the visual appearance of the space.

The new emphasis on recreation also prompted other changes in the landscape. Evidence of former industrial activity was often erased with quarries being refilled. Structures might also be adapted or constructed to facilitate recreational activity. These could be permanent (as in the case of the Grandstand built on Newcastle's Town Moor in 1800 and rebuilt in 1844) or more temporary (like the tower erected at the southern end of the Steine in Brighton from which musicians could entertain promenaders). Even on Bristol Downs, where the 1861 Act seriously hindered any building, the conversion of the windmill at Clifton Camp into an observatory with a camera obscura was designed to encourage recreational pursuits – and constitutes a very visual representation of the move from industry to recreation.

The shift away from earlier uses and towards recreation also had implications for users, and though often framed in positive terms, such as expanding the use of the site "for all", this was not always true in practice. According to Eleanor Straughton, advocates of improvement often claimed that commoners did not make good or effective use of the land.[57] Enclosure commissioners were able to cite examples of people, without common rights, stealing wood and turf from the common or turning out animals when they had no right to do so.[58] There are examples from Durdham Down of the misuse of turf, and Howkins documents similar examples elsewhere. Similarly in the Mousehold case, it was argued that the quarrying undertaken by Pockthorpe brickmakers made the Heath aesthetically unappealing and even dangerous for recreational users. Part of the issue here, as Howkins makes clear, was ambiguity over the use of these spaces. Legally, only those who had property in specific places could exercise rights on the common. Yet in practice, locals – especially the poor – had made use of the commons without any legal justification for doing so. This could include grazing animals (especially ponies, horses and donkeys used for work) and taking wood, bracken or other resources, including gravel and clay. In addition, travellers and vagrants might use the commons as a temporary or more permanent place to stay. Howkins argues that the Commons

Preservation Society was not keen to support these groups and that the process of enclosure in the late nineteenth century often obstructed the opportunities for these members of the community to have a say.[59]

There are also long-standing concerns that because of the location and (in)accessibility of certain commons, the recreational opportunities on offer were more open to some members of the community than others. This concern has been openly expressed about Clifton and Durdham Down. The location of the Downs close to wealthier areas of the city (Clifton, Sneyd Park) makes them ideal recreational spaces for those local residents, but less accessible for those from other poorer districts. One local interviewed for this project explained that getting to the Downs from the poorer part of the city in which they lived requires two separate bus journeys because "'the buses that go to Clifton don't go to the poor places, deliberately" and concluded that the Downs are exclusive "most people can't get there".[60] Similarly, while Newcastle's Town Moor border areas of the city housing a wide range of socio-economic groups, there was concern in the nineteenth century that as a recreational space it was more open to the residents of wealthy Jesmond than those of some of the city's poorer districts.[61]

The emphasis on recreation for all could also hide attempts at social manipulation. The campaigns to convert urban commons such as Town Moor and Mousehold Heath into people's parks in the second half of the nineteenth century were partly about controlling access to those spaces and policing behaviour within them, and often came with a host of new bye-laws about what could and could not be done in those spaces. As Katrina Navickas and Mark Gorman have argued, this was often about reducing the space available for political protests, but it was also connected to broader political agendas, particularly those designed to appeal to the growing middle class.[62] For example, recreational pursuits associated with the lower classes, such as boxing or cock fighting, which had long taken place on urban commons, were suddenly deemed unacceptable. When Norwich City Council took possession of Mousehold Heath in 1866, a long list of bye-laws introduced new restrictions, including strict bans on fighting, bad language and disorderly persons.[63] The Ranger was ordered by the Conservators to post up copies of the bye-laws as well as being instructed to act against men reported to be "prowling about" on the Heath "for grossly improper purposes".[64]

While this partly reflects "Victorian" morals, similar attitudes and innovations can be identified in mainland Europe. By the 1860s, the Warande, in Antwerp, was a problematic green space for the city authorities, which led to calls for gentrification that were similar to the English movement for "people's parks".[65] Yet here, as in England, there was continued intermingling of diverse users. The activities of users of urban commons were far less controllable than in an urban park where spaces were more regulated, gates could be locked at the end of the day and a park warden was employed to watch over the behaviour of users.

New tensions continued to surface after 1900. As Peter Clark has indicated, the twentieth century "saw powerful bureaucratic momentum in the development of green space".[66] From the early twentieth century, green space became less vital as a space that the urban masses travelled to and more important for crowded suburban zones, just as Octavia Hill predicted in 1876.[67] Furthermore, with the expansion of road networks, commons were increasingly threatened by planners seeking better traffic flows for commuters in the mid-to-late twentieth century.[68] Sharp rises in house prices in the late twentieth and early twenty-first centuries have also impacted both positively and negatively on urban commons. On the one hand, middle-class residents are stimulated to protect green spaces that contribute to the economic value of their homes, but this also serves to freeze out other users, such as lower wage earners or minority groups, who may feel that those spaces are not for them.

New tensions have also been generated by the growing recognition of the environmental benefits of green spaces in terms of, for example: carbon capture; offering habitats for wildlife, including rare flora and fauna; and providing drainage that can help to offset urban flooding. Laudable as these new concerns undoubtedly are, they can conflict both with recreational pursuits and with the more traditional uses of urban commons. In Bristol, fencing has had to be erected on the Downs to protect some of the endangered wildflower species from runners. In Newcastle, environmental campaigners lament the monoculture created by grazing and call for the rewilding of parts of the Town Moor. The issues are complex and difficult to resolve. Moreover, different environmental concerns can even conflict with each other. In Norwich, attempts to recreate the historic heathland have been criticised by those who argue that more tree planting would bring greater environmental benefits.

Conclusion

Most of the urban commons in England today have existed as open green spaces for centuries. Moreover, as landscapes that embody the interaction between humankind and the natural environment, they are representative of the long history of the complex relationship between people and place. It is equally clear that the surface continuity of these spaces hides a deeper story of change and adaptation. Their extent and layout, their appearance, the uses to which they are put and the values they are deemed to embody have all been subject to change over time. This should inspire confidence that urban commons can continue to adapt to new circumstances, thereby ensuring their preservation for many generations to come.

Yet it is equally clear that in the past, it was often the most marginalised and vulnerable in society who were negatively impacted by the changing shape, appearance and use of urban commons. This included members of local communities who suffered through loss of employment or means of subsistence, but also less visible users – such as travellers, vagrants and other transient groups – whose members relied on the commons for sustenance and shelter. Moreover, compensation for the loss of these spaces or restricted access to them was rarely sufficient. It was acknowledged as early as the eighteenth century that financial gain to those with common rights was skewed in favour of large landowners.[69] By the late nineteenth century, campaigners such as Octavia Hill highlighted the broader loss to all users of green spaces.[70] For Hill, the implications of loss went far beyond landowners and cottagers, and the provision of allotments could never compensate for the loss of all unappropriated land.[71]

Moreover, the loss was not simply a matter of economics. The identity – as well as the well-being and survival of the inhabitants – was closely tied to these spaces. Medieval and early modern commons were vast working environments that connected local inhabitants to a place. Those people were invested in the land and may not have known much of the wider landscape beyond it.[72] It is clear that dramatic shifts in previously accessible landscapes have been costly to human experience – too often landscapes with which people were intimately familiar changed shape or were lost altogether generally without their consultation and with little recompense. Today, when "efficiency" seems to trump all other values, especially with regard to public money and resources, it is crucial that we pay particular attention to preserving these valuable spaces and ensuring that they are accessible to all citizens regardless of wealth or power.[73]

Rather than simply lamenting past losses, though, we must learn from these past experiences and use the historical resources that we have in the present to protect urban commons for the future. A careful reading of the historical record (both documentary and in the landscape) reveals that in order to secure that future, it is vital to articulate the value of urban commons in ways that make sense to those in power while at the same time preserving their openness, both as physical spaces and metaphorically to a myriad of uses, interpretations and values. Only in that way can we ensure that these now rare and valuable spaces are available for new – perhaps as yet unforeseen uses – for many generations to come.

Notes

1 Elizabeth M. Halcrow, 'The Town Moor of Newcastle Upon Tyne', *Archaeologia Aeliana* XXXI (1953): 149.
2 Kate Ashbrook, *Saving Open Spaces: The Campaign for Public Rights to Enjoy Commons, Green Spaces and Paths* (Pitkin, 2015). Available at: https://www.oss.org.uk/about-us/our-history/saving-open-spaces-catalyst-for-the-commons-preservation-society/
3 Though note Sara Birtles's argument that, in Norfolk at least, commons as registered under the 1965 Act are by no means simply the vestiges of the commons of the past. Sara Birtles, 'The Impact of Commons Registration: A Norfolk Study', *Landscape History* 20:1 (1998): 83.
4 Guy Standing, *Plunder of the Commons: A Manifesto for Sharing Public Wealth* (Harmondsworth: Pelican, 2019), p. 67.
5 I am drawing here on the summary provided by Alun Howkins, 'The use and abuse of the English commons 1845–1914' *History Workshop Journal* 78 (2014): 107–132, which itself draws on Ross Wordie's 1983 work.
6 Howkins, 'The use and abuse of the English Commons', 115.
7 Llewellynn Jewitt, *The Ballads and Songs of Derbyshire* (London: Bemrose and Sons, 1867), pp. 187 and 193.
8 http://hansard.millbanksystems.com/commons/1833/feb/21/public-health
9 ART. XV. – *Report from the Select Committee on Public Walks.* Parl. Papers, 1833, reprinted in *The Westminster Review*, April 1834, p. 500.
10 Ibid.
11 Ibid., p. 507.
12 Ibid., p. 513.

13 'Health and Morals of the People: Enclosure of Waste Lands', *Spectator*, 13 August 1836, p. 15. The article was also reproduced in the *Hereford Times*, 20 August 1836 (Issue CCXVII). Available at: www.britishnewspaperarchive.co.uk (Accessed 26 May 2020).
14 Ibid.
15 Howkins, 'The use and abuse of the English commons', p. 110.
16 Ben Cowell, 'The Commons Preservation Society and the Campaign for Berkhamsted Common, 1866–7', *Rural History* 13:2 (2002): 148.
17 Ibid., p. 148.
18 Mark Gorman, *Saving the People's Forest: Open Spaces, Enclosure and Popular Protest in Mid-Victorian London* (Hatfield: Hatfield University Press, 2021).
19 Jewitt, *The Ballads and Songs of Derbyshire*, p. 187.
20 *The Freemen of Newcastle upon Tyne and the Town Moor*, 50th Anniversary Revised edition (Newcastle, 1999), p. 15.
21 Norfolk Record Office: COL 9/74, Communication from Mr John Wagstaffe upon the inclosure of Mousehold Heath - August 31st 1792.
22 https://britishlistedbuildings.co.uk/101051828-lazar-house-norwich-sewell-ward#.Y4XrUS-l2Tc
23 National Monuments Record: HER 5896. Map reference: NZ 26 NW 33. https://www.twsitelines.info/smr/5896; https://www.heritage.norfolk.gov.uk/record-details?MNF631-Lazar-House-Library-and-former-leper-hospital
24 https://historic-hospitals.com/category/english-hospitals/isolation-hospitals/
25 *Newcastle Journal*, 1 September 1866.
26 Steven Glover, *The History, Gazetteer, and Directory of the County of Derby (Part II)* (Derby: Henry Mozley and Son, 1829), p. 444.
27 HER 5891. Map reference NZ26NW.
28 https://co-curate.ncl.ac.uk/prisoner-of-war-camp-nuns-moor-park/
29 NHER 51904. See https://www.heritage.norfolk.gov.uk/record-details?MNF57545-Area-of-World-War-One-and-Two-practice-trench-digging-and-possible-prisoner-of-war-camp-on-Mousehold-heath
30 Norfolk Record Office: MC 2386/1, 962X6.
31 Norfolk Record Office: MC 2386/1, 962X6 – Letters received by Mrs Statham of Norwich whilst Hans Dittrich was in Norwich, 18 December 1947–7 January 1948.
32 Eneas Mackenzie, *Historical Account of Newcastle-Upon-Tyne Including the Borough of Gateshead* (Newcastle, 1827), pp. 709–14. *British History Online* http://www.british-history.ac.uk/no-series/newcastle-historical-account/pp709-714 (Accessed 20 November 2019).
33 https://www.heritage.norfolk.gov.uk/record-details?MNF26434-Britannia-Barracks-now-part-of-HM-Prison-Norwich
34 https://historicengland.org.uk/listing/the-list/list-entry/1282215?section=official-list-entry
35 Victoria Bates, 'Casson, Elizabeth (1881–1954), medical practitioner', *Oxford Dictionary of National Biography*. Available at: https://www-oxforddnb-com.libproxy.ncl.ac.uk/view/10.1093/ref:odnb/9780198614128.001.0001/odnb-9780198614128-e-63767 (Accessed 16 June 2022).
36 'Charles Barry (1823–1900)', *Oxford Dictionary of National Biography*.
37 https://www.parliament.uk/about/living-heritage/transformingsociety/transportcomm/roadsrail/overview/turnpikestolls/
38 (James Murray), *The Contest. Being an Account of the Matter in Dispute Between the Magistrates and Burgesses, And an Examination of the Merit and Conduct of the Candidates in the present Election for Newcastle upon Tyne* (1774), p. 22.
39 'Projected Railway across the Town Moor', *Newcastle Journal*, 9 December 1837.

40 *The Western Daily Press,* 1 January 1868, accessed 6 August 2020, www.britishnewspaperarchive.co.uk
41 In light of the continued lack of public support for Baker's Road, the committee had no choice but to reject Robinson and Edwards's proposal. *Western Daily Press*, 31 July 1875. Available at: www.britishnewspaperarchive.co.uk (Accessed 13 August 2020).
42 Mousehold Conservators Minutes, 27 May 1886.
43 Mousehold Conservators Minutes for 1886.
44 Ibid.
45 Ibid.
46 Clifton and Durdham Downs: A Landscape History. Final Report, February 2006, p. 22.
47 'Newcastle Town Moor Intakes', *Newcastle Courant*, 22 September 1876; 'Letting of Intakes on the Town Moor', *Newcastle Courant*, 8 September 1876.
48 Mark Bowden, Graham Brown, and Nicky Smith, *An Archaeology of Town Commons in England 'A Very Fair Field Indeed'* (Swindon: English Heritage, 2009), p. 37.
49 Clifton and Durdham Downs: A Landscape History Final Report, February 2006, p. 3.
50 'A View of Brighton. From the Albion Hotel. Old Steine', 'The Old Steine, Brighton'.
51 Tyne & Wear Archives: MD/NC/2/6, p. 24.
52 Tyne & Wear Archives: MD/NC/2/6, p. 96.
53 Tyne & Wear Archives: GU/CW/2/3 - meeting for 2 October 1775.
54 https://www.bristolwater.co.uk/about-us/our-story/our-history (Accessed 29 June 2022); Bristol RO 34901/206.
55 For a detailed account of the Pockthorpe case, see Chapter 4.
56 Bowden, Brown, and Smith, *An Archaeology of Town Commons in England*, pp. 36–37.
57 Howkins, 'The use and abuse of the English Commons', p. 116.
58 Howkins, 'The Use and Abuse of the English Commons', pp. 120–122.
59 Ibid.
60 Interview with FB.
61 'The proposed recreation ground on the Newcastle Moor. Meeting at Elswick', *Morpeth Herald*, 28 May 1881.
62 Katrina Navickas, 'From open space to public space: the idea of the right to air and recreation' talk delivered at the Wastes and Strays event 'Experiences of Urban Green Spaces 1: Historic Perspective of the Urban Commons'; Gorman, *Saving the People's Forest.*
63 *City of Norwich Act, 1867. Draft Scheme with Respect to Mousehold Heath*, within Judgement delivered on the case of the City of Norwich versus Browne.
64 Mousehold Conservator Minutes, 1886.
65 Bart Tritsmans, 'Versatile Green: An Alternative Perspective on Urban Green Space in Late Nineteenth-Century Antwerp', *Urban History* 42:1 (2015): 95–99.
66 Peter Clark, *The European City and Green Space: London, Stockholm, Helsinki and St Petersburg, 1850–2000* (London: Routledge, 2016), pp. 17–19.
67 Octavia Hill, *Our Common Land and Other Short Essays* (Cambridge: Cambridge University Press, 2011), pp. 188–189.
68 Clark, *The European* City, pp. 17–19.
69 *Wagstaffe upon the inclosure of Mousehold Heath – August 31st 1792* (COL 9/74), NRO, Communication from Mr John.
70 Hill, *Our Common Land*, pp. 191–195.
71 Ibid.
72 Michael Hough, *Out of Place: Restoring Identity to the Regional Landscape* (London: Yale University Press, 1990), p. 35.
73 For a forcefully argued account of the threat posed to commons by the rise of Friedrich August von Hayek's vision of a free market economy, see Standing, *Plunder of the Commons*.

7 Conflict and the Contested Common

Introduction

Urban commons have a long history of association with conflict and contestation. From Kett's Rebellion in the sixteenth century through to the Black Lives Matter campaigns of the twenty-first century, there are numerous documented examples of urban commons as protest sites. Conflict could also erupt spontaneously as large groups of people gathered in these spaces for recreational purposes. Commons could themselves be the focus of contestation – the objects as well as the setting for conflict. Certain urban commons would undoubtedly have been lost if key groups had not contested plans to encroach upon and develop the sites. Contestation has also arisen over the enacting of common rights in those spaces, over the appropriate use of them and over how they should be managed and by whom.

In recent years, scholars have become interested in the locations in which conflict and contestation took place and in the significance of space as a determining factor in the cause or unfolding of such events. These issues have figured in work on popular protest including that by Andy Wood on the early modern period and by scholars such as John Barrell, James Epstein, Christina Parolin, Katrina Navickas and Steve Poole on the late eighteenth and early nineteenth centuries. These historians have touched on urban commons in the course of their work, with Wood writing about Kett's Rebellion and the camp on Mousehold Heath, Navickas exploring the role that urban commons in the north of England played in the Chartist and Reform movements and Poole examining conflict in Bristol, including that relating to Clifton Down.[1] Yet these studies tend to be limited both chronologically and geographically, and the focus is specifically on protest – and the urban commons as sites of protest – rather than on the broader relationship between urban commons and contestation and the ways in which these spaces might invite conflict.

That deeper relationship has sometimes been explored in individual studies of specific urban commons and the conflicts that secured their preservation. Examples here include Neil MacMaster's excellent article on Mousehold Heath and Mark Gorman's book on Epping Forest.[2] These studies are, of course, important and have informed what follows, but their focused nature makes it difficult to use them to draw more general conclusions.

DOI: 10.4324/9781003204558-9

There are also one or two studies which have adopted a longer-term case study approach centred around key themes. Briony McDonagh and Carl J. Griffin's article "Occupy! Historical geographies of property, protest and the commons, 1500–1850" is a good example. It centres on occupation as a strategy of protest and draws on a range of examples.[3] McDonagh and Griffin make an important argument about the multifaceted nature of occupation as a strategy for resisting enclosure – that it was not only a symbolic gesture but also a means of reclaiming and remaking the space for the public. That argument will be referenced below, since similar strategies are in evidence in our case studies, but the article focuses on rural rather than urban commons. In adopting a collaborative approach and comparing four very different urban commons, the Wastes and Strays project has made it possible to generate some more general conclusions about the deep and complex relationship between urban commons and contestation.

This chapter argues that the prevalence of conflict in and over urban commons is a direct consequence of their peculiar character. In various respects, urban commons are hybrid spaces. Most obviously they are neither wholly urban nor wholly rural, but an in-between space where the rules of neither the town nor the countryside can be consistently applied. We might also think about them in the light of the tripartite model of space set out by Henri Lefebvre and Edward Soja.[4] On this account, "First space" is the perceived or material space as it exists in reality. "Second space" is the more abstract dimension of space as an imagined, symbolic or representative entity. "Third space" involves a combination of the two and refers to the lived experience of the space by those who inhabit or pass through it in everyday life. Thinking about space in this way opens the possibility of counterspaces involving resistance to dominant uses, constructions and codification.

In the case of urban commons, these possibilities are heightened by three characteristics that have rendered urban commons susceptible to conflict and contestation. In the first place, they are **Empty Spaces** – the "First space" element is often relatively weak since the spaces are open and uncluttered by buildings and other infrastructure. This offers potential for multiple individuals and groups within society to inhabit, use and imagine the space in their own ways. Second, they are **Enduring Spaces** – which have often existed as commons for centuries. In this respect they are a remnant of the past in the present, often containing evidence of past uses within the landscape. Of course, as Chapter 6 demonstrates, their endurance is often more myth than reality and changes to the size and shape as well as the uses of urban commons have occurred over time. Moreover, conflict has sometimes emerged because the legal frameworks required to protect those spaces are necessarily fixed at a particular moment in time, which can sometimes be a barrier to adaptation as new circumstances arise. Together, the spatial and temporal characteristics of urban commons mean that they are above all **Enigmatic Spaces** – ambivalent in their character and amenable to being subject simultaneously to a range of varied and sometimes conflicting interpretations. Moreover, each urban common has its own history of competing uses both at particular points in time and over years and even centuries.

Empty Spaces

Urban commons today are generally open – perhaps even apparently wild – spaces located in the midst of cities. They therefore provide a contrast to their surroundings which tend to be more built up and therefore, on the surface at least, more ordered. Moreover, the activities that take place on urban commons are generally of a temporary nature, often leaving little or no trace in the landscape.

Owing to these characteristics, urban commons have been used throughout history for communal gatherings. While some of these events were peaceful both in their planning and execution, others degenerated into conflict or were deliberately contestatory from the outset. A sense of this variety can be gained by considering a handful of specific examples, starting with two very different gatherings that took place on Newcastle's Town Moor in the mid-nineteenth century just under 20 years apart.

The morning of 21 April 1848 was cold but dry, a welcome relief from the persistent rain of previous weeks.[5] Although it was Good Friday, the placards displayed on the Moor were not of a religious character, but instead declared "High Treason against the Sovereign People", accusing the Government of hurrying through a "Gagging Bill" ("The Security of the Crown and Government Bill") to prevent people from speaking freely on matters of reform.[6] Soon after midday, confusion descended since the hour announced for the commencement of proceedings had passed but none of the intended speakers had arrived. Despite this, the crowd remained calm, the speakers eventually appeared and the proceedings began. Soon after, the rain began to fall once more, meaning that banners could not be displayed nor musical accompaniment provided. The conditions, however, failed to dampen the spirits of the crowd. After several speeches, it was suggested that the meeting be adjourned until Monday afternoon due to the inclement weather, but the proposal was dismissed by cries of "go on" from the crowd. A number of speeches were enthusiastically given and positively received by as many as 3,000 gathered on the Moor, all of whom then peacefully dispersed, a sympathetic observer commenting that "not a single policeman was to be seen on the Moor" that day.[7]

The second example comes from Race Wednesday on 27 June 1866.[8] The final race had just been run and P. C. Anderson had left his position at the Newcastle turn of the race course to return to the police tent and retire for the day. As he did so he spotted a large crowd running from the top of the hill down towards the tents. As the crowd reached him, one of its number (later identified as Thomas Findlay) struck the policeman on the head with a heavy stick. Despite the blow, P. C. Anderson managed to get hold of his assailant, wrench the stick from him and convey him (with the help of several civilians) to the police station where he was locked up. While this was going on, a second man, Patrick Kelly, also struck the policeman. In all, there were thought to be as many as 200 people in the crowd. Witnesses claimed that some of those involved had declared themselves to be Fenians and that one had been heard shouting: "To hell with Garibaldi and save the Pope". In the end, 18 men from across the region were charged with being involved in a riot.

The first example, though an organised political event, was a pregnant conflict. There is no evidence of any violence occurring, of participants coming to blows or of a stand-off between the crowd and the authorities. The whole event was conducted (if the newspaper account is to be believed) in a dignified and civilised manner. Yet the background to this Chartist meeting was more contentious, given the events that had taken place in Paris in February and in other European cities in the following weeks and months. In this context, the meetings like the one described above which took place across the country in April 1848, often on urban commons, can be viewed as a call to arms.

By contrast, the second example was both unexpected (at least as far as the authorities were concerned) and violent. It erupted during a recreational gathering which brought a crowd of locals onto the Moor. While there is evidence to suggest at least some of those involved were motivated by political concerns, the whole affair was less ordered, the objectives of the protagonists less clear and the conflict may simply have erupted spontaneously out of an atmosphere made volatile by sport and alcohol.

Another two examples, this time from Brighton's Valley Gardens, reveal that similar activities could be viewed very differently, depending on the identity of the instigators or participants. Fireworks were deployed as part of the celebrations that accompanied the first visit of the Prince of Wales to the town in 1782. They were said to have been "let off to the east of the royal Duke's house ... which was then an open space" to allow the royal visitor "a distinct and pleasing view".[9] Thirty-five years later, individuals found letting off fireworks on the Old Steine were arrested. On 4 November 1817, a public notice was issued warning against the holding of Bonfire Night celebrations. Despite this, a group composed mainly of young boys went ahead with their plans and at 9 p.m. a tar barrel appeared. The boys were arrested and the barrel captured and extinguished, but this led to rioting in the town with attacks made on the High Constable's House and other properties. The militia eventually restored order, but only after several people had been killed and injured.

As these examples make clear, conflict on urban commons could take a variety of forms. It could be highly organised or spontaneous, peaceful or violent, explicitly political or not. Despite this last point, many of the cases of conflict on urban commons had a political edge to them, and it is worth considering in more depth the role of urban commons in overtly political conflict: the prevalence of such gatherings, the types of issues involved and the purposes they were designed to serve.

Urban Commons as Spaces of Protest

The open and empty nature of urban commons and their location close to urban centres have long made them obvious locations for protests involving large groups of people. Organised gatherings for political purposes took place in all four of our case study locations and over a remarkably long period of time. When Robert Kett and his followers gathered on Mousehold Heath in July 1549, they were said to have chosen a location that had been used by earlier protestors at least as far back as the Peasants Revolt in 1381. At the other end of the spectrum, Valley Gardens in Brighton has regularly been used for demonstrations in the twenty-first century, including a vigil for the murdered woman Sarah Everard in March 2021 and Extinction Rebellion protests against climate change two months later.

Political reform meetings were a particularly common form of public gathering on urban commons during the nineteenth century. As early as August 1819, a group assembled on Newcastle's Town Moor to complain about the lack of fair representation in Parliament and to call for reform, and in October of that year, a meeting was held headed by local radicals Eneas Mackenzie and John Marshall "to consider the late proceedings at Manchester".[10] Chartist and later Reform meetings were held on the Town Moor, Mousehold Heath and Bristol's Clifton Down, and the fountain on the Steine in Brighton was a site for political speeches from the time of the first Reform Bill.[11]

Meetings on more specific or localised political issues were also held in these spaces. In the 1710s, public meetings were held on Clifton Down in defiance of the Whig Corporation. These were used by Jacobites as an opportunity to express their hostility to the Hanoverian regime. In October 1872, there was a large gathering of mainly Irish residents of Tyneside (many of whom worked in local factories) on the Town Moor to protest at the holding of Irish political prisoners by the English authorities and to call for their release.[12] Activists might also take advantage of other events for their own purposes. In 1913, Mary Richardson of the Clifton branch of the Women's Political and Social Union took the opportunity of King George V's visit to the Downs for the Agricultural Show to deliver a piece of paper promoting the suffragist cause directly to the King. Having dropped a scroll of paper promoting the cause onto the King's lap, she was immediately attacked by the crowd, one spectator slapping her and another hitting her over the head with an umbrella before she was taken into police custody.[13]

Gatherings were also held to highlight economic grievances. These often involved groups of workers linked to local trades. Wool-combers camped on Mousehold Heath during a strike in 1752 and weavers returned repeatedly to the Heath in the 1820s. In Newcastle, pitmen took to the Town Moor in October 1850.[14] Meetings of this kind continued to be held into the twentieth century. A rally on Clifton Down in 1915 was designed as a protest against the dangers facing railway workers and in 1921 a mass meeting was held there on the theme of unemployment.

These political and economic gatherings served multiple ends. In the first place, they provided an opportunity for co-ordinated action and the presentation of a united front. Many resulted in a collective document through the mass signing of a petition or voting on a plan of action. Second, they afforded participants an opportunity to demonstrate their numerical power to the authorities. Newspaper accounts often noted the numbers involved and groups sometimes played on this strength. In July 1822, for example, the Norfolk weavers, having drawn up a petition to their employers complaining of their low wages, threatened to start a riot in Norwich if their demands were not heard.[15] Third, they were a means of publicising the cause among local residents. To this end, the meeting on the common was often preceded or followed by a procession through the town. The aim was to be seen – and heard – by as many people as possible. Katrina Navickas has highlighted the significance of crowds transforming the soundscape as well as the appearance of the space through the use of music and shouts, boos, cheers and other forms of what she calls "aural graffiti".[16] Brass bands would often accompany the procession; rosettes, sashes or favours in particular colours would be worn; and banners bearing pithy slogans would be used to highlight the cause. The 3,000 weavers who met on Mousehold in 1827 marched to Norwich Guildhall to address the city magistrates playing music and waving placards complaining about the reduction in their pay.[17] A newspaper account of the pitmen's demonstration on the Town Moor noted that its purpose was "to inform the public of their grievances", which included poor ventilation in mines and the lack of an education and fair pay for mine workers.[18]

There were both practical and symbolic reasons for holding these gatherings on urban commons. As well as the size and location of these spaces making them ideal for accommodating large groups of people, the liminal character of the space meant that they were often seen as being beyond the direct surveillance of the authorities. They also became associated with protest through repeated use for this purpose over centuries, and by the mid-nineteenth century, they were a recognised venue for political meetings.[19] Moreover, by gathering on common land, protesters could enact as well as demand their rights. This can be seen in the case of Kett's rebellion, in which one of the rights they claimed was the right to "comon uppon the Comons".[20] Similarly, McDonagh and Griffin emphasise that in the case of the Digger movement of the seventeenth century, the fact that they were reclaiming common land that they believed had been stolen from the people was significant.[21] By the nineteenth century, the right to gather was closely associated with the right to free speech. Consequently, restricting access to these spaces, or diminishing them through enclosure, could be interpreted as an assault on people's rights and freedoms and as a deliberate means of denying them political participation.[22] In this context, gathering and protesting in these spaces constituted both the voicing and the enactment of rights.

Controlling and Codifying Urban Commons

Just as protestors used urban commons as a venue for their demonstrations, so the authorities sought to codify the space themselves – placing their mark upon it and emphasising to the public that they retained power and control over these spaces. One way in which this was achieved was through the use of urban commons for public executions. We can find many examples of individuals being executed on urban commons.[23] In some cases, the choice of location reflected where the crime itself had taken place. In 1548, a man named Cond was hung in chains on Durdham Down on the spot where he had murdered his master. Almost 200 years later, in 1743, Captain Maccartny, who had been found guilty of murdering a man named Beechy on Bristol Downs, was hung from the ravine. As a result, the ravine became known for a time as Maccartny's Gully.[24] John Latimer refers to Andrew Bennet (or Burnet), Henry Payne and a sailor named Abseny who were also hung in chains near to the Sea Wall so that their bodies would be seen by passing vessels. Bennet and Payne robbed and murdered a coachman travelling from Bristol to Stoke Bishop and Abseny was found guilty of the murder of a 13-year-old servant girl. The Bristol authorities regularly displayed criminals who had committed robberies on the Downs at the gallows on Clifton Down, which was located at the north end of Pembroke Road (formerly Gallows Acre Lane). It was last used in 1783 to hang Shenkin Protheroe for the murder of a drover on the Downs, but the body was cut down and buried after rumours circulated that he was descending from the gallows at night to haunt the streets of Clifton.[25] These very public executions were a way for the authorities to mount a warning to other would-be offenders, but they were perhaps also intended as a means for them to re-establish their control over an empty, liminal – even lawless – space.

It was not only those who had committed an offence on the common who met their end there. Urban commons were often a regular location for the public execution of felons who had committed crimes at other locations within the city. Gloucestershire's county gibbet pole was close to Durdham Down and at least 15 individuals were executed on Newcastle's Town Moor in the late eighteenth and early nineteenth centuries.[26] Once again, there were pragmatic reasons for the choice of location, not least that the spaces could accommodate a large crowd and their proximity to the city, but the symbolic resonance was no doubt also a factor. The status of commons as marginal land usefully reinforced the marginality of the criminals being punished. Moreover, public executions were a good way for the authorities to display their power and stamp their meaning on the space, thereby serving as a warning to those (such as political protesters) who might wish to use the space for their own ends.[27] In this regard, it is interesting to note the similarities in the ways in which executions and political demonstrations were conducted. At both, the element of spectacle was key. Hangings were often preceded by a procession in which the unfortunate felon was led publicly through the streets to draw both the crime and the punishment to the attention of the public. As in McCartney's case, the body might then be left in situ to remind passers-by of the punishment. In other cases, the empty gibbet could serve a similar purpose.

Executions were one way in which the authorities could codify urban commons, sending a message to the public about the nature of the space, the Corporation's control over it and appropriate behaviour within it. Another means of achieving this end was through displays of military power and strength. From as early as the seventeenth century, urban commons were used as a rallying point for soldiers. Again, the expansive and open nature of the space and their proximity to urban centres made them ideal for this purpose, but here too a symbolic element was involved.

Bristol Downs was repeatedly used for the gathering and training of troops from the time of the British civil wars in the 1640s. Newcastle's Town Moor was used as a military camp during the 1745 Jacobite campaign.[28] Military drills were clearly also taking place on Mousehold Heath in the mid-eighteenth century, since there was a report of a man dying after sustaining a bullet wound to the arm while watching these activities in 1758. These drills were perhaps the source of the lead balls and shot found in the area of the Long Valley.[29] A militia camp was established in Brighton in 1784 and following this, military parades were held regularly in the area of Valley Gardens and the Old Steine. This practice seems to have continued into the nineteenth century. A print from 1847 shows the Steine being used for military parading and a source from 1864 suggests that military bands performed there daily.[30] Military reviews and sham fights were also held regularly on Newcastle's Town

Moor and on Bristol Downs in the late nineteenth century. It is possible that these displays of military power were linked more directly to reform agitation, since in Newcastle the first military review on the Town Moor took place in July 1868, meaning that it may have been a deliberate response to the reform agitation of the previous year. By filling a space that had recently been used for political demonstrations with over 4,000 armed troops, the authorities were perhaps seeking to make a point.

Of course, the problem for the demonstrators and the authorities alike was that the events they organised were, by their very nature, temporary and left little if any mark on the landscape. From an early date, however, urban commons were also used for more permanent military and punitive establishments, which would have been a more imposing – and perhaps therefore more effective – means of codifying the space.

As early as 1558, the Lords of the Manor of Brighthelmston granted a parcel of land to the town authorities to build an armoury between Ship Street and Black Lion Street close to the Old Steine. From the nineteenth century, far more of these establishments were built, often being located on the common itself. In 1805, rare encroachment was made onto Newcastle's Town Moor for a military barracks. By 1863, a shooting stand had been erected on Mousehold Heath and 20 years later land was obtained for the Britannia Barracks. Around the same time, in 1885, the Home Office obtained a portion of land on Mousehold Heath for the building of Norwich Prison, with further land being transferred for an extension in 1960.

As the case study chapters indicate, all four of our case study areas were used extensively during the two world wars for a variety of military purposes, including as airfields, military training facilities and defensive structures. In Bristol, there is evidence of some retaliation by locals to this display of military force. Peace protests were staged on the Downs in 1932–1933 and again in 1937–1938. Speakers in 1933 explicitly condemned not just the expenditure on arms but also the attempt to create a "war spirit" through military display.

As these examples demonstrate, the fact that urban commons were open – even empty – spaces led protestors and the authorities alike repeatedly to use the spaces for their own purposes in an attempt to fill, control and even codify the space. The building of barracks and prisons on urban commons was one means by which the authorities sought more permanently to fix their codification. Another, more subtle means of doing so was by turning various urban commons into "People's Parks".

People's Parks

The earliest recorded public park on an urban common was Moorfields in London, which was converted in the autumn of 1607. It boasted two walks bordered by walls and trees which crossed in the centre where people could "take the ayre". It was, however, in the nineteenth century that the conversion of urban commons into parks was adopted more widely and became a controversial topic. The arguments in favour of the establishment of so-called "People's Parks" in urban areas arose out of the debates in the 1830s concerning the health and well-being of the rising numbers of urban workers.[31] Given their location and nature, the country's remaining urban commons were obvious candidates for transformation into People's Parks. In fact, plans for the "improvement" of these spaces were already being voiced in the late eighteenth and early nineteenth centuries. John Wagstaffe's 1792 proposal for the enclosure of Mousehold Heath included a proposal for the ornamentation of that part of the heath that was closest to the city with tree planting and the creation of vistas and avenues.[32]

In Newcastle, a reader of the *Newcastle Guardian and Tyne Mercury*, who called himself a "Lover of Improvement", was one of the first to call for the conversion of parts of the Town Moor and Castle Leazes into public parks.[33] Like Wagstaffe, this Tyneside commentator believed that this required a transformation of the space. He noted that while the Moor and Leazes were open to the public, many people did not take up the opportunity to use them for recreation. In particular, he claimed, "Many females are deterred by the cattle, and the want of proper roads". Later that year, it was suggested that the idea be discussed by the Corporation. Other towns were providing new parks and walks for their inhabitants and it was noted that Newcastle was behind on this, but it was the late 1850s and early 1860s before the issue was properly picked up again.

In August 1861, the Town Moor Committee of the Town Council put forward a scheme for the future management of the Moor. As part of this, it was recommended that an application be made to Parliament "to render the Leazes and portions of the Moor ornamental ground" with "the formation of rides, drives, cricket and drill grounds – a gymnasium, in fact, for all classes".[34] While the plan found favour in some quarters, it provoked a hostile response elsewhere. The *Newcastle Journal* saw the move as the latest attempt by the Corporation to steal control of the Town Moor from the Freemen and traced the history of that conflict right back to the 1770s. It was not the health benefits of access to open space that was in question: the "pure and bracing air of the Town Moor and Leazes" were deemed to have "always been highly conducive to the health and enjoyment of the inhabitants of the city" and local doctors were said to have often recommended that their patients walk on the Moor "for the restoration of [their] health". Rather, the issue was whether anything actually needed to change: "the Town Moor, Nun's Moor and Castle Leazes already form the most healthy and extensive public park in the north of England, and perhaps in the kingdom". It would therefore be a waste of taxpayers' money to engage in the expense of turning it into a formal park which would then "confine the public to certain prescribed portions of an extensive estate over which they can now roam and engage in various sports at pleasure".[35]

Those calling for a People's Park were not silenced by these objections and the idea continued to be mooted throughout the 1860s into the 1870s. Objections too continued to be voiced. In February 1865, a Mr Muras made reference to the scheme for the conversion of the Leazes into a park. Had this occurred, he asserted:

> there would have been only some six or eight feet of footpath to walk upon, and if any of their children had ever attempted to play at cricket, as they at present did, on the green sward, the gentlemen in blue would soon be upon them, and walk them off perhaps to gaol.[36]

This sense that the opportunities for residents would be restricted rather than expanded by this particular "improvement" was a common concern. At a meeting of Elswick ratepayers in late 1869, it was again asserted that setting aside a portion of the Town Moor or Leazes for recreation would amount to a restriction of the privileges the people currently possessed, since "The whole Moor and Leazes were open to the public, and might be enjoyed by them".[37] In the end, a compromise was reached. Instead of the whole area of the Town Moor, Leazes and Nuns Moor being "improved" in this way, part of Castle Leazes was separated off to form Leazes Park, which opened in 1873.

In Norwich too, the idea of creating a People's Park played into more localised debates about the fate of Mousehold Heath. Though Wagstaffe's proposal was not implemented, by the mid-nineteenth century Mousehold was an increasingly popular place for members of the middle and upper classes to walk and exercise. These users expressed their unhappiness about the pits and quarries that they saw as disfiguring the landscape. The town council debated this issue on 19 May 1857 and when the Dean and Chapter offered Mousehold to the city in December 1864, they did so on the condition that it be transformed into a "People's Park". A special People's Park Committee was established in 1876 and met regularly for the next six years. The character and use of the Heath after the Corporation finally took control reflected its status as a People's Park. For example, new bye-laws were introduced at that time, designed to lay down what constituted acceptable behaviour within the space and to establish a procedure for the enforcement of those rules. In this respect, the authorities strengthened their codification of the space.

Enduring Spaces

Urban commons are distinctive not only in extending over space, but also over time. Despite their diversity in other respects, surviving English urban commons generally have a recorded history that extends back to early modern or medieval times. It is, then, part of the nature of urban commons that they are enduring spaces that have been in existence over a long period of time.

Legal Frameworks

The preservation of these spaces has depended on distinctive historical and legal frameworks specific to the individual place and circumstances. Bristol Downs is the only one of our four case studies to be protected as a registered common, with common grazing rights recorded on the commons register in 1965. Newcastle's Town Moor has, since the late eighteenth century, been protected by specific Acts of Parliament (1774 and 1988) and has been preserved largely as a result of the rights of the Freemen that were enshrined in those Acts. Mousehold Heath also owes its survival in part to parliamentary legislation, in this case the City of Norwich Mousehold Heath Scheme Confirmation Act 1884, and to the commitment of the body established by that Act – the Mousehold Heath Conservators. Their role in managing the Heath was reinforced under the City of Norwich Act 1984. Today, environmental concerns, the status of part of the area as a Site of Special Scientific

Interest and the presence of scheduled ancient monuments are key to the area's continued survival and maintenance. Brighton's Valley Gardens has no formal legal protection, but instead depends on Brighton City Council for its protection. The designation of the area as a Conservation Area provides some security.

While crucial in securing the continued existence of these spaces, the systems employed have generated various problems and complications. In part, this is because there is often more than one body involved: Bristol City Council and the Merchant Adventurers in the case of Clifton and Durdham Downs; Newcastle City Council and the Freemen for Town Moor; and Norwich City Council and the Conservators (and also the Defenders) in the case of Mousehold. This sharing of responsibility can itself be a source of disagreement and contestation. In addition, each of these legal frameworks was established at a particular point in time and reflects the circumstances at that moment. In many cases, the provisions were not well designed to move and flex with the changing times.

Changes Over Time

Although by the mid-nineteenth century the main function of urban commons was shifting away from agriculture and towards recreation, grazing rights continued to be seen by some as crucial to the preservation of these spaces, which could be a source of conflict. In the mid-nineteenth century, when Bristol City Council sought to secure Durdham Down against development, the method they adopted was to purchase the few remaining tenements that came with the right to graze sheep on the Downs and then exercise those rights.[38] Similarly, by the second half of the nineteenth century, the cattle that grazed on the Town Moor were no longer essential to the work of the town's artisans and merchants as they had been back in medieval times. Moreover, even the more recent justification that the grazing of cattle on the Moor was a means of providing residents of the town with cheap milk could no longer be sustained. At a public meeting of the burgesses of East All Saints' Ward, convened to consider the idea of converting the Town Moor into a public park, Alderman William Lockey Harle pointed out that the value and purpose of the Town Moor was different in his own time from what it had been a century earlier:

> It might be all very well in the last century when the cows provided milk for the town to so restrict the moor, but that had all passed away. We now considered the Town Moor not merely for the grazing of the cows, but as a positive enjoyment for the people of Newcastle. Country milk was preferred now, and that was amply supplied by the agricultural population.[39]

On this basis, Harle went on to argue that "the whole system of feeding the cows was an antiquated absurdity" which should end. Harle kept up his campaign. Almost ten years later, he was still concerned that cattle were hindering the improvement of the area:

> The Moor was no longer useful or valuable for the purposes of the town; and poverty and rinderpest had reduced the 600 cows it used to feed to 300

> half-starved animals, which every agriculturalist ridiculed. The whole system of pasturage on the Moor, as well as the intakes and fencing, was barbarous, and the ridicule of agriculturists.[40]

Harle's was not, however, the predominant view. His proposal was rejected and cows continue to graze on the Town Moor today.

On Mousehold Heath, it was not the grazing rights that were the issue, but rather a different set of "historic common rights" associated with quarrying. As described in Chapter 3, the residents of nearby Pockthorpe had been extracting gravel, sand and clay from the Heath for the purposes of brickmaking since at least the sixteenth century. By the nineteenth century, this activity came under attack by the authorities and wealthier residents of Norwich on the grounds that it impacted on the aesthetics of the space and hindered recreational activity. One local resident complained in 1855 that a walk on the Heath might be "abruptly terminated by an involuntary somersault into one of the numerous gravel pits of the domain".[41]

In their campaign against the authorities, the Pockthorpe residents based their defence in part on their use of the land for sand and gravel extraction "from time immemorial". Yet, in the end, their claim was not upheld since the right of estover – to take produce of the soil such as wood, peat or minerals – was only to be exercised by local householders to support their own subsistence, whereas the Pockthorpe residents were using the materials they took from the Heath for commercial purposes. Interestingly, when the conflict was finally resolved in the 1880s, grazing rights were explicitly mentioned, in part to underline the rejection of other historic common rights. In July 1882, the Corporation asserted that the only lawful common rights that still applied to Mousehold were rights of pasturage for sheep and cattle levant and couchant. This was asserted, even though grazing on the Heath had declined significantly by this point. Moreover, it is interesting to note that unlike on Bristol Downs, where grazing rights are still exercised in order to protect the space, the grazing rights on Mousehold were not registered under the Commons Registration Act 1965, meaning that the Heath ceased to be a "common" in the narrow legal sense soon after. Since that time, the protection of Mousehold Heath has been determined less by reference to historic rights of the past and instead on the basis of the protection of the natural environment for future generations.

Conflict Over Encroachment

Conflict could also arise over enclosure and encroachment when the terms of the legal codes that protected urban commons became outdated or ceased to be properly enforced. The Newcastle Town Moor Act of 1774 had been explicitly designed as a means of resolving problems associated with the letting out of portions of the common. Yet, by the mid-nineteenth century, conflicting interpretations of the details of this Act were generating further contestation.

The interpretation of the Act already appears to have been at issue by the autumn of 1852, since a map was drawn up to show the area in which the Freemen had to live in order to graze cows on the Moor or benefit from the rents generated and the areas of land that had been leased out for cultivation in accordance with the 1774 Act.[42] It was, however, during the 1860s, amid the debates about converting the Moor into a People's Park that the disputes became heated. In September 1861, a meeting of the Freemen and Stewards concerned with deciding the annual letting of plots descended into technical legal debate. The solicitor and Freemen E. Story declared that on his reading of the Act of Parliament, the body of stewards had the power to let portions of the surface of the Town Moor for a period of seven years and that "no piece of ground could be re-let until the whole of the Town Moor and Castle Leazes had been let".[43] Others argued that for many years, the leases that had been granted had not been strictly in line with the letter of the Act.[44] By 1868, Mr Samuel Rowell, who frequently spoke on behalf of the Freemen, was admitting that "it was impossible literally to comply with the Act, and that all they could do was to carry out its spirit".[45] This conflict had wider repercussions. It led the Freemen into a dispute with the Newcastle Grand Stand Company over its organisation of the popular annual race meeting, resulting in 1868 in popular action to tear down the fences. Consequently, the following year, the Company took the unprecedented measure of threatening to cancel the races.[46]

In Bristol, too, debates over encroachment hinged on uncertainty over past legal arrangements. A report produced on behalf of the Society of Merchant Venturers in 1859 concluded that arrangements made in the 1780s and 1790s between the Society and Samuel Worrall (town clerk of Bristol between 1787 and 1819) and his heirs had resulted in a relinquishment of rights to the land by the Society and its trustees. Worrall and his partner Thomas Sims (a House Carpenter from Bristol) owned a 12-acre plot of rocky ground known as Shortgrove, lying open and unenclosed on Durdham Down.[47] The Society of Merchant Venturers also owned a piece of land in fee trust of approximately 14 acres, also known as Shortgrove and also lying open and unenclosed on the Downs adjoining Worrall and Sims's land, but which was wooded rather than rocky. In July 1787, Worrall and Sims alerted the Society of their intention to divide and enclose their property.[48] As the ground lay adjacent to Society holdings, they wanted to establish a boundary so that future disputes could be avoided. The correspondence between the two parties demonstrates the difficulties encountered in reaching an agreement based on historic indentures that were mostly text descriptions of land assets. However, Worrall and Sims were able to establish their right to ground on Durdham Down, including a lime kiln and quarry that were marked out on the De Wilstar Survey of 1746.[49]

The Merchant Venturers were keen to ensure that parts of their property did not become alienated from the larger mass of the Downs, so they proposed an exchange. In return for Worrall and Sims's land to the north of Upper Belgrave Road, the Society would grant as much as was necessary of their piece of ground called Shortgrove to the south of the road. Worrall and Sims agreed, but insisted on quantity in lieu of quality because of the poor condition of the Society's landholding.[50] A boundary trench with stone markers was to be set out to distinguish the two pieces of land, and a small strip of land was kept by the Society between the new boundary and Upper Belgrave Road. The terms of the agreement were recorded in a Deed dated 31 May 1788. However, the following year, Worrall and Sims measured the exchanged piece of land and realised that it was less than 12 acres – not the original 16 they had agreed to – and refused to go on with the exchange until the Society rectified the mistake.[51]

A memorandum dated 30 June 1791 was attached to the 1788 Deed, which stipulated that the Society of Merchant Venturers would add two additional portions of ground so that the exchange could take place. The 1788 Deed and the 1791 memorandum were endorsed by the Society, attested by the Master of the Society and the Clerk and were to be enrolled in the High Court of Chancery. The documents specifically stated that the land was *the gift of the Society* and that both portions of land should henceforth and forever be held together by Worrall and Sims and their heirs.[52] However, when the actions of Worrall's heir were questioned in 1859 and the documents re-examined, they were found to be unsigned.[53]

Despite this, the 1859 report found in favour of the Worrall family because Worrall had been in possession of the piece of land from at least 1791 without the title being questioned by the Society. In fact, the Merchant Venturers had expressly recognised the agreement as late as 1844 by maintaining the boundary between Worrall's property and the Downs. Further, Worrall's heirs were able to demonstrate that the family had continuously managed the upkeep of the property as landholders and let various parts of the land, including the quarry. The Society was therefore deemed to have no grounds to dispute the original 1788 and 1791 documents on the basis that they had not been signed, and the Worrall family were within their rights to enclose their property with a wall.[54] The 1859 case also made a point of questioning the limits to which "commonable rights" could be used to prevent enclosure. While few properties with rights of common remained in the parish of Clifton, the case tested the legitimacy of those with common rights in Henbury – could they destroy the enclosure walls on the basis that their cattle had previously had access to the unenclosed land? However, as a wrangle common provided a form of legitimate trespass rather than a right, the commoners of Durdham Down in Henbury had no claim and as the original transfer had preserved a piece of land of superior quality for the commoners, there were no grounds to suppose that the commonable rights over the Shortgrove land had been diminished.[55]

Roads as Encroachment

Attempts to build roads across urban commons – to respond to changing communication needs – were also a source of contention. In Bristol, the erection of tollgates on Stoke Road and Westbury Road that led across the Downs sparked fierce opposition in the early eighteenth century. Between 1726 and 1746, rioters repeatedly tore down or burned the toll gates that were erected and subsequently repaired. The objection was that the tolls were levied on wagons, horses, cattle and sheep when passage had originally been free and that this created a burden for the local people. In 1726, four of the rioters were captured, but because there was no fixed penalty for destroying tollgates, they were released. During the 1730s, every gate between Bristol and Gloucester was said to have been destroyed with rioters collecting money from travellers and local landowners and farmers encouraging the destruction by providing the rioters with alcohol. By the 1740s, the trustees of Clifton and Durdham Down were unable to keep up the repair of the tollgates and the money collected was insufficient to repair the turnpikes.[56]

Though the Morpeth and Ponteland Turnpikes constructed across Newcastle's Town Moor proved less controversial, the road constructed by William Ord between the Ponteland Turnpike and his Fenham estate did lead to conflict. After Newcastle Council had overruled the Freemen and allowed Ord to build a road rather than just a bridleway, a man named "Whitby" was accused by Ord of breaking up the new road. Ord brought a lawsuit against Whitby, which was heard at King's Bench and resulted in victory for Ord.

Roads over Bristol Downs continued to be a problem in the nineteenth century. The road constructed by William Baker in 1862 to provide convenient access for the residents of Sneyd Park provoked such an outcry that having approved the venture the Downs Committee was forced into a U-turn and had to re-turf the road, though it was subsequently relaid as a footpath in 1881.

Roads continue to provoke issues on urban commons in the twenty-first century, not least in Brighton. In 2018, a new three-phase plan for the regeneration of Valley Gardens was initiated. Central to the plan was the improvement of traffic circulation around the area and the enhancement of the public spaces. The problems of traffic congestion and pollution around this green space are long-standing issues. Also controversial was the construction of the urban motorway across part of Newcastle's Town Moor in the 1970s.

The Return to "Traditional" Landscapes

The conflicts over the building of roads across urban commons reflect the obstruction that these traditional spaces could present to modern needs – in this case, communication needs. Yet, in recent years, conflicts have also emerged over calls to return urban commons to their "traditional" state. This has been a particular issue in Norwich. One of the aims of the current management plan is to "re-heath" part of the area, cutting down trees and shrubs so as to restore Mousehold Heath to past glory. As the oral history testimonies revealed, this strategy has found favour with some residents but has provoked anger from others. Part of the issue concerns what is understood by the "traditional" or "original" state of the common. The natural state of Mousehold is not healthland, but woodland. It was only the extensive grazing of animals on the land from the eleventh century that prevented the regrowth of trees and shrubs creating the Heath. Thus some complain that the policy to "re-heath" the area will bring about a return not to Mousehold's original state, but rather to a later man-made bucolic idyll. Moreover, some campaigners also argue that this is unacceptable at a time when we are seeking to increase rather than reduce the number of trees in the country to address the climate crisis. Of course, heathland brings its own environmental benefits, so the issues are not straightforward.

This very recent conflict over an urban common reminds us that while urban commons may be Enduring Spaces and remnants of the past in the present, the relationship between past, present and future is neither straightforward nor uncontentious.

Enigmatic Spaces

As a result of both their open and their enduring nature, urban commons are inherently enigmatic spaces. The activities that take place on them are predominantly temporary, meaning that they can be occupied by different groups at different times. At the same time, the legal frameworks that govern them are open to interpretation and can quickly become outdated, leaving considerable ambiguity as to how the space can or should be used. As a result of this, conflicts have frequently emerged between competing individuals and groups who have different visions of the urban common and seek to use it for different – sometimes incompatible – purposes.

Agricultural Improvement versus Common Rights

One long-standing tension around the use of common land was that between agricultural improvement and common rights. This was already in evidence at the time of Kett's Rebellion in the sixteenth century. At the heart of this conflict was the disregarding of common rights by landowners in the interests of agricultural improvement. While Mousehold Heath was not one of the commons that was under threat, the decision of the rebels to camp there was no doubt significant, and in doing so they were not only arguing for the right to "comon uppon the Comon" but were also simultaneously enacting that right. Elsewhere, this reassertion of common rights was more directly provocative. McDonagh and Griffin note that opposition to enclosure in the sixteenth century often involved the breaking down or burning of fences and hedges that had been erected around the enclosed land. Alternatively, gates might be opened or animals led back onto the land constituting a reassertion of common grazing rights.[57] Similar strategies were employed by the Newcastle Freemen in their conflict with the Corporation in the late eighteenth century. For example, after Joshua Hopper had been granted the lease to 89 acres at the west end of the Moor in January 1772, on the condition that he improve the land using appropriate agricultural methods, the land was enclosed and a house built. Unhappy that the enclosure had been imposed upon them by the Corporation and that they were not involved in the management of the scheme, a group of Freemen, including Nathaniel Bayles and Henry Gibson, from the surgeon's guild demolished the fence and set fire to the gates and allowed their cattle back onto the enclosed land, thereby challenging the decision that had been made and reasserting their common rights.[58] According to one supporter of the Freemen, the fear was that more enclosures would be made for the benefit of individuals and at the expense of the poor Freemen and widows who would be "stripped" of their benefits.[59]

Similarly, in 1848, a motion was passed by the Stewards Committee that the crop of grass on the intakes of the Town Moor be sold for hay and the proceeds be divided between improvements to the Moor and relief for poor burgesses and widows. However, some Freemen were opposed to the decision and, as the *Newcastle Journal* reported, they "thought proper to remove the gates on the intake at the Fenham Ends, the best portion of the crop, and drive their cows on the grass, thereby damaging the crop to such an extent, as to render it almost unsaleable for hay; the resolution of the stewards, was, therefore, abandoned, and the notice of sale publicly countermanded".[60]

The Rise of Recreation

More commonly, conflicts arose because the growing emphasis on common land as a space for recreation from the late eighteenth century conflicted with its traditional domestic, agricultural and industrial uses. These conflicts are particularly evident in Brighton and Bristol where the development of local spa resorts impacted on the space.

At Brighton, work on the area around the Pavilion in the early nineteenth century necessitated the removal of a large barn belonging to a Mr Howell, which had been used to house animals that grazed in the area. A wooden weigh scale, which was originally situated at the bottom of North Street, was also removed at the Prince's request. It had been used to assess tolls on products such as hay or bricks entering the town via the London road. According to Alderman Henry Martin, in 1871 the scales had to be removed because they were large and ugly, which suggests they were impacting on the aesthetics of this popular pleasure garden.

In Bristol, the creation of the Hotwells Spa marked a moment of transformation in the space immediately generating conflict in relation to earlier uses. When they began the development, the Merchant Venturers immediately alienated the local population by charging them for pump water at Hotwells, which they had previously regarded as a right. Trouble also emerged around the mining of "Bristol Diamonds" along the Avon Gorge for sale to spa visitors. As George William Manby commented in 1806:

> These rocks are the property of that opulent and respectable society of Merchant-Venturers of Bristol, who will not allow this beautiful mass to be defaced. It is to be lamented their interference has not restricted the prodigious havoc that is daily making on its neighbouring rock [Durdham Down], as the venerable majesty of this truly sublime wonder of nature is receiving daily insult and robbed of some ancient grace by the rude hand of mercenary labour.[59]

By 1840, complaints voiced in the newspapers suggest that the exploitation of the rocks was excessive: "frightful inroads have been made upon those bold projecting rocks that once overhung the river". By 1845, workmen were busy at between six and eight different sites along the rock face and were using explosives to remove the quartz which was both noisy and dangerous.

The picture is, however, complex since the new development also brought with it new domestic and industrial uses of the space, which, in turn, came into conflict with its recreational functions. In the late nineteenth century, the space beside Clifton Down Road opposite the end of Northcote Road was often used to beat the carpets from the large houses of Clifton. Large wooden posts were erected from which carpets could be hung to be beaten. In May 1864, the Downs Committee approved the erection of four additional posts for the purpose and in June 1871, they agreed that the posts should be repaired.[61] However, in the 1870s and 1880s, various complaints were made to the Committee about the nuisance caused by the carpet beating and the cleaning of mattresses that also took place there. Initially, the Downs Committee allowed the practice to continue, insisting that they were unable to identify a better location. In 1885, however, the former mayor of Bristol and local businessman Christopher J. Thomas made a complaint, which prompted the issuing of a notice that carpet beating on the Downs should cease from 30 June 1885. The posts that had been used for the purpose were then removed. In July 1885, a deputation representing local carpet beaters appeared to have secured a compromise involving restricting the times at which carpets could be beaten, but at the next Town Council meeting, the Downs Committee was urged to frame new bye-laws which included a ban on carpet beating and an attempt to rescind the ban was lost by three votes to four. Even then, it took time for the new rules to be implemented. The proposed bye-laws (which also banned the hanging and drying of clothes, the erecting of tents, flower picking and rock collecting) were approved by the Merchant Venturers in April 1890 and an amended draft was considered by the Council the following December. The new bye-laws finally came into effect on 1 August 1892 and notice boards stating the relevant bye-law were erected in the area in which carpets had been beaten.[62] The Ranger was then instructed to take the names and addresses of any offenders, and in 1893 he presented 20 such names to the Committee.[63]

The Steine, at the southernmost tip of Valley Gardens, was also used for domestic activities that required extensive outdoor space, including drying clothes and cleaning carpets. It was also used for jobs associated with the fishing industry. From at least the sixteenth century, the Steine was used for drying nets as well as for boat construction, repair and storage. It was perhaps the practice of using the Steine for the construction and repair of boats that led to it being used as a more general workspace for town residents, where vehicles of all kinds could be repaired. By the eighteenth century, there is also evidence to suggest that the area was used for coal storage and as a venue from which merchants could sell their wares.

These activities are depicted in the visual sources, which suggest that different and sometimes competing activities were taking place on the Steine simultaneously. John Donowell's *A Perspective View of the Steyne at Brighthelmstone* (1778) (Figure 7.1) shows that the area was already being presented as a fashionable location for elite recreation by the last quarter of the eighteenth century.

Yet other images suggest that older activities had not ceased and there was conflict in the 1770s between the owners of lodging houses around the Steine who wanted it turned into a formal promenade site and the fishermen who used it to dry their nets, who did not.[64] In the end, the fishermen lost out. By 1778, railings enclosed the southern part of the Steine and new footpaths and turf was laid. Within

Figure 7.1 A Perspective View of the Steyne, at Brighthelmston in the year 1778 from the South End.

the next decade, the whole of the area was re-turfed and enclosed into five sections using wooden railings. The pictorial evidence from the 1790s shows sheep grazing and boat storage both taking place within the fenced area of the Steine. Similarly, *View on the Steine, Brighton* (1808) depicts workers in the foreground cleaning a carpet in close proximity to the parading urban elites, and an image from 1819 shows washing lines for drying clothes stretched out across Valley Gardens close to St Peter's Church.

In Bristol, there are also examples of tensions between military and recreational uses. From as early as 1819, the Clifton Cricket Club had the approval of the Lords of the Manor of Henbury to use a piece of land close to the Sea Wall for their matches. The same area was used by the Duke of Beaufort's Regiment of Yeomanry for their annual training exercises. The damage the military activities caused to the turf resulted in cricketers sustaining injuries. Consequently, in 1849, the Clifton Cricket Club was allowed to move to a different area of the Downs that was not used by the Yeomanry. In 1941, the golf course was put out of action when part of it was requisitioned for the military vehicle park.

Different kinds of recreational pursuit could also conflict with each other. As recreation became the dominant use of the Downs, the range of sporting activities taking place in that area increased dramatically, prompting various complaints to the Downs Committee.[65] Opponents of the Clifton Cricket Club complained about the crowds attracted by the popularity of the game and the damage caused to the turf by carriages being drawn near to the pitch (1888); that tents that were erected to provide refreshments and shelter for players and spectators had not been taken down when they should have been (1895); about the timing and method of grass cutting (1920s); and about the use of a portable coke boiler for making refreshments (1926). In the end, the Club seems to have tired of the conflict, and in 1927 they used a different pitch for the year, letting theirs to two other cricket clubs. They returned to the Downs in 1928, but in April 1930 they informed the Downs Committee that they were leaving the Downs for good.

Golf proved equally controversial. In December 1876, an application was made to the Downs Committee to establish a Clifton Golf Club, but it was refused. In May 1889, a letter of complaint was sent to the Headmaster of Clifton College on account of golf holes having been made on the Downs without permission, but later that year, the Downs Committee agreed to establish a nine-hole course on either side of Ladies Mile. Even so, trouble continued. In 1893, Mr Herbert Ashman, the future Lord Mayor of Bristol, gave notice of a motion prohibiting golf on the Downs. In the end, a compromise was reached that there would be no golf after 2 pm. The Downs Committee regularly received complaints about near misses from golf balls and there were also complaints by golfers that sheep got in the way of play. Golf ceased on the Downs during the war and today bye-laws explicitly prohibit the driving, chipping or pitching of hard golf balls on the Downs.

Conclusion

This chapter has explored the theme of conflict and contestation on urban commons. It has argued that throughout history, these spaces have been a location for, as well as a source of, contestation and that the prevalence of conflict on urban commons owes much to their nature. Their status as Empty Spaces in the midst of increasing urban sprawl made them a popular location for temporary gatherings organised by local people and by the authorities. This meant that different groups of people could project different (and sometimes conflicting) visions onto the space and that none of them (not even the authorities) were able to permanently codify the space in their terms. Codifications were always open to resistance, challenge and replacement. This contributed to the characterisation of urban commons as Enigmatic Spaces that were used and understood differently by different stakeholders. As a result, they operated in a state of constant flux, ambiguity and possibility.

The distinctive spatial features of urban commons must be set alongside their special temporal character. Those that have survived into the twenty-first century have proved their status as Enduring Spaces over a long period of time. Most have remained open spaces continuously since medieval times. Yet, despite that longevity changes in size and shape, landscape and use have occurred (sometimes arising out of or generating conflict). Moreover, the legal frameworks that have ensured the survival of these spaces themselves often emerged out of conflict, but also often went on to generate further conflict, particularly when their strictures no longer fitted the reality of changing circumstances. Thus, this too contributed to the sense of urban commons as Enigmatic Spaces. The legal frameworks themselves have proved open to interpretation and have engendered ambiguity.

The histories of our four case studies have shown that conflict has sometimes had a positive impact. Resistance to encroachment has been essential to their survival at certain points in time and the variety of visions regarding the value of these spaces has itself been a strength. Yet conflict can also be destructive, alienating certain groups, hindering the activities of others and even damaging the space itself. The task for the future is to find a way of allowing, protecting and celebrating the Enigmatic character of these spaces, so that they can remain Empty and Enduring for generations to come.

Notes

1 Andy Wood, *Riot, Rebellion and Popular Politics in Early Modern England* (Basingstoke: Palgrave, 2002); Christina Parolin, *Radical Spaces: Fenues of Popular Politics in London, 1790-c.1845* (Canberra: ANU Press, 2010); James Epstein, *Radical Expression: Political Language, Ritual, and Symbol in England, 1790–1850* (London: Breviary Stuff, 2014); Katrina Navickas, *Protest and the Politics of Space and Place, 1789–1848* (Manchester: Manchester University Press, 2016); Katrina Navickas, 'From Open Space to Public Space: The Idea of the Right to Air and Recreation', paper delivered at the webinar 'Experiences of Urban Green Spaces: What Is an Urban Common?' I, 13 April 2021; Steve Poole and Nicholas Rogers, *Bristol From Below: Law, Authority and Protest in a Georgian City* (Woodbridge, NJ: The Boydell Press, 2017).

2 Neil MacMaster, 'The Battle for Mousehold Heath 1857–1884 "Popular Politics" and the Victorian Public Park', *Past and Present* 127 (1990): 117–154; Mark Gorman, *Saving the People's Forest: Open Spaces, enclosure and popular protest in mid-Victorian London* (Hatfield: University of Hertfordshire Press, 2021). See also Robert Allen, 'The Battle for the Common: Politics and Populism in Mid-Victorian Kentish London', *Social History* 22:1 (1997): 61–77.

3 Briony McDonagh and Carl J. Griffin, 'Occupy! Historical Geographies of Property, Protest and the Commons, 1500–1850', *Journal of Historical Geography* 53 (2016): 1–10.

4 Navickas, *Protest and the Politics of Space and Place*, pp. 14–15; Katrina Navickas, 'Thirdspace? Historians and the Spatial Turn, with a Case Study of Political Graffiti in Late Eighteenth- and Early Nineteenth-Century England', in *Spatial Cultures: Towards a New Social Morphology of Cities Past and Present*, ed. Sam Griffiths and Alexander von Lünen (London and New York: Routledge, 2018), pp. 67–75.

5 Fergus O'Connor comments on the recent bad weather in his speech in the Commons 18 April 1848. Hansard, vol. 98 cc453–79. Avaliable at: https://api.parliament.uk/historic-hansard/commons/1848/apr/18/crown-and-government-security-bill#S3V0098P0_18480418_HOC_4

6 O'Connor used that term to refer to the bill in Parliament. Ibid.

7 'Chartist Meeting on Newcastle Town Moor', *Newcastle Guardian and Tyne Mercury*, 22 April 1848.

8 This account is based on 'The Late Fenian Riot on the Town Moor', *Newcastle Journal*, 6 July 1866.

9 Richard Sickelmore, *History of Brighton and its Environs, From the Earliest Period to the Present Time*, 5th edition (Brighton: C. and R. Sickelmore, 1827), pp. 33–34.

10 *Radical Monday: A letter from Bob in Gotham to his cousin Bob in the country, containing an account of that glorious day!!..* (Newcastle, 1819). The University of Manchester, Peterloo Collection R228953. Reprinted as a postscript to,'A Slight Dusting of Whigs at the Newcastle Meeting', *Black Dwarf*, 4:5, 9 February 1820), p. 180; Eneas Mackenzie, *Historical Account of Newcastle-Upon-Tyne Including the Borough of Gateshead* (Newcastle upon Tyne, 1827), pp. 81–83.

11 *Newcastle Guardian and Tyne Mercury, 22 April 1848*; Tyne & Wear Archives Service: DF. COW/C/1743; *Norfolk Chronicle and Norwich Gazette*, 2 May 1840; Erredge, *The Ancient and Modern History of Brighton*, p. 271.
12 *Morpeth Herald*, 2 November 1872.
13 'Suffragette and King', *Tamworth Herald*, 12 July 1913. Available at: www.britishnewspaperarchive.co.uk (Accessed April 21, 2020).
14 'Great Meeting of Pitmen on Newcastle Town Moor', *Newcastle Guardian and Tyne Mercury*, 19 October 1850.
15 *The Bury and Norwich Post*, 17 July 1822; *Jackson's Oxford Journal*, 12 February 1827; and *The Bury and Norwich Post,* 16 December 1829. All Available at: www.britishnewspaperarchive.co.uk (Accessed 20 December 20, 2019).
16 Navickas, 'Thirdspace? Historians and the spatial turn', in *Spatial Cultures*, ed. Griffiths and von Lünen, p. 74.
17 *Jackson's Oxford Journal*, 17 February 1827.
18 Great Meeting of Pitmen on Newcastle Town Moor', *Newcastle Guardian and Tyne Mercury*, 19 October 1850.
19 Navickas,'Thirdspace? Historians and the spatial turn', p. 74
20 'Kett's demands being in rebellion', 1549, reprinted in Anthony Fletcher and Diarmaid MacCulloch, *Tudor Rebellions*, 5th edition (London and New York: Routledge, 2014), pp. 156–159.
21 McDonagh and Griffin, 'Occupy!', on the Digger movement more generally see John Gurney, *Brave Community: The Digger Movement in the English Revolution* (Manchester: Manchester University Press, 2007).
22 Gorman, *Saving the People's Forest*, pp. 42–43 and 144–145.
23 Sarah Collins, 'A Dangerously Empty Space: Crime and Punishment on English Wastelands', *Landscapes*, 23:1 (2022): 48–66.
24 John Latimer, *The Annals of Bristol in the Nineteenth Century*, Volume 1 (Bristol: W&F Morgan, 1887), p. 104.
25 Alfred Edmund Hudd, *Proceedings of the Clifton Antiquarian Club*, Volume 5 (Clifton: Clifton Antiquarian Club, 1904), p. 93; Fells, *The Little Book of Bristol*, chap. 12.
26 Poole and Rogers, *Bristol from Below*, p. 167; *Newcastle Journal*, 28 May 1859.
27 *Newcastle Journal*, 28 May 1859.
28 *The Journal of the Reverend John Wesley* (New York, 1855), Volume 1, pp. 351–355.
29 NHER 53875, NWHCM 1973.296, NWHCM 1974.546.1.
30 John Erredge, *The Ancient and Modern History of Brighton* (Brighton: W. J. Smith, 1867), p. 194.
31 For more detail on this see Chapter 6.
32 Norfolk Record Office NCR COL 9/74, Communication from John Wagstaffe upon the inclosure of Mousehold Heath, 31 August 1792.
33 Letter to the Newcastle Guardian and Tyne Mercury, 20 February 1847.
34 'The Moor and Leazes, *Newcastle Journal*, 21 November 1862.
35 'The Corporation and the Town Moor', *Newcastle Journal*, 20 September, 1861.
36 February 1865.
37 'The Town Moor Question', *Newcastle Guardian and Tyne Mercury*, 4 December 1869.
38 Richard Goldthorpe, *Clifton and Durdham Downs: A Landscape History Final Report* (Bristol: Bristol City Council, 2006), p. 3.
39 'Newcastle Town Moor Question', *Newcastle Couran*t, 21 September 1877.
40 'Newcastle Town Council', *Newcastle Courant*, 20 November 1868.
41 Cited in Whyte, 'Remembering Mousehold Heath', p. 41.
42 'Newcastle Freemen's Guild', *Guardian and Tyne Mercury*, 16 October 1852.
43 'Town Moor Intakes', *Newcastle Journal*, 5 September 1861; 'The Town Moor Intakes – Protest of Freemen', *Guardian and Tyne Mercury*, 7 September 1861.
44 *Newcastle Guardian and Tyne Mercury*, 7 September 1861.

45 'Letting of the Town Moor Intakes', *Newcastle Guardian and Tyne Mercury*, 5 September 1768.
46 *Newcastle Guardian and Tyne Mercury*, 9 January 1869.
47 Bristol Archives: SMV/6/5/6/5, Report of Law Officers regarding Durdham Down Enclosure, 4 May 1859.
48 Bristol Archives: SMV/2/4/2/28, Letter from Samuel Worrall and Thomas Sims to the Society of Merchant Venturers, July 13, 1787.
49 Bristol Archives: SMV/2/4/2/28, Letter from Samuel Worrall to Mr Osborne attorney, March 20, 1788.
50 Bristol Archives:SMV/2/4/2/28, Letter from Samuel Worrall to Mr Osborne, July 11, 1788.
51 Bristol Archives, SMV2/4/2/28, Letter from Samuel Worrall to Mr Osborne, May 30, 1789.
52 Ibid.
53 Ibid.
54 Bristol Archives: SMV/6/5/6/5, Report of Law Officers regarding Durdham Down Enclosure, 4 May 1859.
55 Ibid.
56 See John Latimer, 'Clifton in 1746', *Clifton Antiquarian Club*, 28 November 1900 (Volume 5), p. 27; Latimer, *The Annals of Bristol*, pp. 155–158.
57 McDonagh and Griffin, 'Occupy!', p. 4.
58 James Murray, *The Contest. Being an Account of the Matter in Dispute Between the Magistrates and Burgesses, And an Examination of the Merit and Conduct of the Candidates in the present Election for Newcastle upon Tyne* (n.p. 1774), p. 25.
59 Ibid.
60 'The Freemen and the Town Moor Intakes', *Newcastle Journal*, 15 July 1848.
61 Gerry Nichols, *Clifton and Durdham Downs: A Place of Public Resort and Recreation* (Bristol: Bristol Branch of the Historical Association), p. 15.
62 Maurice Fells, *The Little Book of Bristol* (Cheltenham: The History Press, 2015), chap. 12.
63 Nichols, *Clifton and Durdham Downs*, p. 16.
64 Sue Berry, 'Pleasure Gardens and Regency Seaside Resorts: Brighton, 1750–1840', *Garden History* 28:2 (2000): 224.
65 Nichols, *Clifton and Durdham Downs*.

Part III

The Future of the Urban Common

8 "Urban" or "Urbane"? Models for Governance and Property Rights in the "Urban" Commons

This chapter will examine different historical models of commons governance and how these may be adapted and developed to better protect and enhance the use value of urban commons for the future. It will also consider the relationship between legal models for governance and property rights in the commons and how these, in turn, shape and reflect conceptions of "belonging", identity and notions of communal ownership and "property" in a contemporary context.

"Property" Rights and the Classification of Urban Commons

The legal classification that has been applied to "property" in the urban commons (and indeed commons more generally) is largely derivative of property rights that reflect the historical function of the commons as a source of natural resources for community use. These reflect the origins of common land in the medieval manorial system. Until comparatively recently, the principal user rights over common land would have been largely agricultural in nature. Historically, there was no right of access allowing members of the public to enter common land, or to use it for recreation or exercise. All common land has an owner, and s/he was entitled at common law to use the land as they wished – as long as their use did not infringe the rights of the owners of common rights to graze livestock on the common or to extract other natural resources such as peat or gravel. Nevertheless, since the mid-nineteenth century, the rights of the owner of the soil have been progressively restricted.

The origins of the property rights over many modern urban commons in this earlier conception of the "common" can be seen in the case studies examined in this work. For example, the right to graze cattle on Town Moor originally gave the freemen of Newcastle a ready supply of milk and other dairy produce to supplement their diet and provide an additional income if needed. Similarly, in the case of Clifton and Durdham Downs, the right to graze sheep on the downs (a right still subsisting over Durdham Downs) was an adjunct of the agricultural resources available to farms and other institutions near the common.

The legal classification of rights and obligations over the land reflects these earlier priorities for land use, which have, of course, changed substantially in the modern era. An important theme that we will elaborate upon below is the failure

DOI: 10.4324/9781003204558-11

of the enduring legal classification of property rights in urban commons to change and reflect their contemporary role in urban society, as reflected in changes in the use made of the space by local communities.

"Property" and "Property Rights" in the Commons

Urban commons today are multidimensional spaces, with multiple (and sometimes competing) land uses that are reflected in a complex web of legal interests, property "rights", customary arrangements and informal socially sanctioned practices. Until comparatively recently, a right of access to common land could only exist as an adjunct of a property right in the land giving access to the common resource (e.g. the rights of the owner of the soil or common rights for gathering turbary or grazing livestock). And these property rights were principally agricultural in nature and reflected the origins of the commons in a mediaeval agricultural economy. This position has today been substantially modified by an overlay of statutory rights, giving members of the public free and open access to common land. Whether the public have a legally sanctioned "right" of access and use will today depend upon (i) the existence of property rights vested in members of the community and/or (ii) the existence of statutory rights granted by an Act of Parliament giving members of the public recreational access to the common, notwithstanding an absence of property rights vested in them.

Although questions of public access to our commons no longer rely solely upon property rights, it remains the case that "property" interests retain an important role in not only defining rights of access but also shaping the way in which the urban commons are managed and by whom. The history and development of property concepts in our urban commons explain and define, for example, their contemporary management and the way it is organised. In the context of the case studies presented in Chapters 2–5, for instance, it explains the systems of "dual" management exemplified by Town Moor in Newcastle and Clifton Downs in Bristol. The role of the Freemen of Newcastle in the management of Town Moor, and now set out in the Town Moor Act 1988, is underpinned by their ownership of the herbage rights over the moor, while freehold ownership is vested in the City Council. Both bodies therefore have important property interests in the land itself, and their joint management (and exercise of these property entitlements) is shaped by the terms of the 1988 Act. Similarly, the dual management arrangement established in the Clifton and Durdham Downs Act 1861 is attributable to the freehold ownership of Clifton Down by the Merchant Venturers and of Durdham Down by the Bristol City Council.

One important question for the future of our urban commons is the extent to which property rights should retain a role in defining and shaping public access and in determining the manner in which a common is to be regulated and managed. Historically, access to the land resource had always been dependent at common law on the existence of a property right. "Property" rights at common law fulfil a resource allocation function, vesting individual elements of resource utility on the holder of the right in question. A common right of grazing, for example, gives its holder the right to take grass by the mouth of grazing livestock.[1] The herbage rights held by the Freemen of Newcastle are a manifestation of this type of "property right", that is, a right that gives named individuals a right to take a defined element of the land's resources (in this case, grass for grazing). But property rights must be held by a named individual or individuals – they cannot be held by a constantly changing group or by the "public" at large.[2] Land ownership in English property law remains founded on the principle of exclusivity of resource use, and this does not include a liberal notion of reasonable access – there is no concept of "quasi-public" land in English law to which a test of reasonable access could be implied.[3] The common law concepts of "property" and "property rights" are grounded in – and reflect – elements of resource utility that are capable of "ownership" and transfer. An inevitable consequence of this approach is that communal or "shared" use of land has never been recognised as creating "property" rights in members of the community. Where common property rights exist (e.g. to graze animals on the common), these are treated as individual property rights in the hands of the appropriator (commoner) and not as a "public" or communal "right" as such. The key distinguishing factor is not, however, the fact that a property right can only subsist if it has an identifiable "owner". Several statutes give members of the "neighbourhood" the right to open-air recreation on privately owned land, for example, on a Town and Village Green.[4] While one is a resident of the relevant neighbourhood, therefore, one has recreational "rights" to use the land. But if you cease to be a local resident, you will in theory cease to enjoy them. It is the *fluctuating* nature of the holders of the interest in the land that prevents it from being recognised as a "property" right at common law.[5]

For this reason, communal rights to use common land have had to be introduced through legislation and are reflected in statutory use right conceptions – not in property rights as such. So, in relation to public access, the principle that elements of common resource use can only be exclusively reflected in property rights is now heavily qualified by the introduction of the "right to roam" by the Countryside and Rights of Way Act 2000, as explained below. The owner of the soil, nevertheless, in principle retains the right to exploitation of the land resource in whatever manner s/he wishes, subject only to the restriction that in doing so, they must not inhibit the exercise of any common rights subsisting over that land or any statutory right of public access. In three of the case studies presented in this work, a statutory right of public access for recreational purposes is conferred by a private Act of Parliament: the Mousehold Heath Scheme Confirmation Act 1884 for Mousehold Heath in Norwich; the Town Moor Act 1988 for Newcastle; and the Clifton and Durdham Downs Act 1861 for the Downs in Bristol. In the absence of a bespoke Act of Parliament conferring public access on a specific urban common, any land that is registered as common land under the Commons Registration Act 1965 or Commons Act 2006 is also automatically considered "access land" under the right to roam legislation.[6]

Although the right of public access to our urban commons – especially for recreational use – is an important policy concern in a modern context, it follows that this is not reflected in the manner in which property rights in the commons are created, organised or protected. This remains primarily concerned with issues of resource allocation, where the focus is on the common as a source of economic resources and not as a provider of cultural ecosystem services like recreation or biodiversity.

Property Rights – the Legacy of the Manorial Model

The origins of the rather idiosyncratic organisation of property rights in common land lies in the manorial system, which vested ownership of the "waste" of a manor in the lord of the manor, while recognising the use rights of the local community. The statutes of Merton (1235) and Westminster II (1285) confirmed the lord of the manor's rights in the soil of the manorial waste, but restricted their freedom of action by recognising the legal validity of the use rights of commoners. The manorial lord, being owner of the soil, retained a wide range of property rights, including mineral and game rights. At the same time, the lord's tenants were allowed to use the manorial waste for grazing their livestock, collecting timber or turf, or many other purposes, in a manner that was recognised by local custom and protected by the local manor courts.

In building upon customary recognition of these practices, English law developed an elaborate classification of commoners' land use rights.[7] Most common rights are "appurtenant", that is, they are attached to a land holding (the "dominant tenement") as a subsidiary right. Some land use rights could also be held independently of a holding of land and are known as common rights "in gross". The common law recognised six categories of common right. These are not exclusive categories, and furthermore, some rights that are encountered in practice are difficult to place within the established categories. The most significant, *common of pasture*, the right to graze livestock on the common was, and remains, very important to the rural economy in upland areas of England and Wales. It was also a feature of several of the case studies of urban commons presented in this work. The management of the collective grazing over large agricultural commons was, prior to the Commons Registration Act 1965, organised around one of two principles in most areas – by the application of the principle of levancy and couchancy or by the imposition of "stints" as a measurement of permitted grazing. The rule of levancy and couchancy determined the number of grazing livestock allowed on the common by reference to the capacity of the land to which the rights attached to feed livestock over the winter months. Stints, on the other hand, permitted a fixed number of livestock to graze the common – a practice still observed, for example, in the allocation of stints to graze cattle on Newcastle Town Moor.[8]

The other common rights most frequently encountered today are common of turbary and common of estovers. These are sometimes referred to as "hearth" rights, in that in principle they must always be attached (appurtenant) to a building or house. *Common of turbary* gives the right to take peat or turf for fuel for heating the dominant tenement, usually a cottage or house. *Common of estovers* is similar and gives the right to take wood or other vegetation for necessary purposes. The common law subdivided this right into three subcategories, which largely reflect its agricultural origins. "House bote" was the right to take timber for fuel or to repair a house; "plough bote" was the right to take timber for making or repairing agricultural implements; and "hay bote" was the right to take timber, shrubs or brush to make or repair hedges and fences. The right also extends to taking animal bedding (e.g. bracken or rushes), gorse, heather and reeds. Other rights of common include the *right of pannage* – the right to graze pigs in woodland or forests – and rights to take fish, wild animals and the soil itself. Rights to take minerals or constituent parts of the soil are recognised as *profits a prendre* in law and can also be the subject matter of a right of common. Most rights to take soil will be appurtenant to land and subject to the same restrictions as turbary and estovers. This was a key point in the nineteenth-century litigation concerning the extraction of clay and soil on Mousehold Heath in Norwich[9] and the development of a nascent brickmaking industry by members of the Pockthorpe community. The High Court refused to recognise a right to take clay for commercial purposes vested in the community, and the case ultimately formed the backdrop to the establishment of the Mousehold Heath Scheme Confirmation Act 1884, discussed further in Chapter 3.[10]

One of the most interesting features of the case studies presented in this work is the manner in which what is now in each case "urban" green space emerged from the disintegration of the manorial system. Importantly, however, in each case it did so in different ways and at a different pace in the different geographical locations. And in relation to the different case studies, this has led to the subsistence of different categories of land use right over each of the urban commons that we have studied. So, Mousehold Heath (Norwich) is an example of a former manorial estate that was invested with the status of a managed urban common in the Victorian period in circumstances drawing heavily on prevailing notions associated with the "people's parks" movement. Both Clifton and Durdham Downs (Bristol) on the other hand remained common land and were registered as such under the Commons Registration Act 1965, although common grazing as a commercial activity and associated agricultural land uses have long since ceased and the grazing rights do not reflect modern land use priorities. Town Moor in Newcastle has been regulated by a private Act of Parliament since the eighteenth century and has a unique status that owes nothing to the manorial system of landownership and land use. Nevertheless, vestiges of the land use rights prevalent in the former manorial system are evident in, for example, the regulation of grazing by cattle on Town Moor through the issuance of "stints". Similarly, Valley Gardens (Brighton) is a green space that has grown out of the land use changes resulting from the expansion and development of Brighton as an urban centre around it.

The Common as Resource: Changing Priorities

Most recreational access to common land has, until comparatively recently, been enjoyed as a permissive licence granted by the landowner – one that gives members of the public no legally enforceable *right* to enter land for the purposes of open-air recreation or sport. The "exclusivity" view of the rights of the owner to bar public access to most commons now needs to be qualified.

Urban Commons – From Inclosure to "Managed" Recreational Space

The gradual introduction of legislation to give the public recreational access rights over common land started in the nineteenth century, with the passage of several local and private Acts of Parliament protecting the more important surviving urban and metropolitan commons, including (but not limited to) several of our case studies; for example, the Epping Forest Act 1878, the Mousehold Heath Scheme Confirmation Act 1884 and the Clifton and Durdham Downs Act 1861. These gave the public rights of recreational access over the specific commons to which they applied and did so on terms reflective of the wider Victorian commons movement led by pioneers like Octavia Hill, Lord Eversley and Robert Hunter. The Victorian legislation marked a progression in public policy from (i) encouraging the inclosure and improvement of common land to (ii) a protectionist stance, preventing further inclosure in order to retain open spaces for recreation and other public uses, especially near rapidly growing urban centres, and finally to (iii) promoting schemes for the management of the surviving urban commons.

The provisions for managed access in our case studies must be seen in historical context, especially against the prevalent policy expressed in Victorian legislation protecting those significant areas of urban common that had survived the inclosure movement. Between 1750 and 1845, the inclosure of common land was effected principally through private Acts of Parliament, and at the height of the inclosure movement, no less than 4,804 private Inclosure Acts enclosed 850,000 hectares of manorial waste and a further 1.82 million hectares of common land, including much of the remaining common pasture in England and Wales.[11] The rapid growth of urban populations during this period meant that areas of common near urban centres were increasingly used for recreation. As noted above, however, the use of common land for recreational public use was not recognised or protected by law. Moreover, increased public use of the remaining commons resulted in tensions between the economic exploitation of the land by commoners – those with legal rights to take the produce of the land – and the wider public seeking recreation on common space near urban centres. Land values increased as the towns and cities expanded and increased recreational pressures on land use meant that the commoners' income was reduced. If a common was enclosed, furthermore, then the commoners received no compensation and the urban population lost their traditional places for recreation.[12]

The trend in legislative policy moved strongly from 1845 towards the preservation of surviving urban open spaces and promoting their management for recreational use by the urban populace. This started with moves towards making inclosure more difficult in the Inclosure Act 1845 itself, which laid down a standard process for all land inclosures that required the inclosure of common land to be effected in future through a complex procedure culminating in the confirmation of a provisional inclosure order by Parliament.[13] The 1845 Act also contained several provisions to protect commons near large towns. No land was to be enclosed without Parliamentary authority, for example, if it was within 15 miles of London or within two miles of any large town or city with a population of 10,000 inhabitants, within 2½ miles of a town or city of 20,000 inhabitants or within three miles of a town or city of 30,000 inhabitants.[14]

Thus, although the 1845 Act in one sense represented the culmination of the inclosure movement, it also signalled a change in public policy. This fundamental change – from promoting the inclosure of common lands to making inclosure more difficult and protecting the remaining common space – was amply demonstrated in successive Acts of Parliament throughout the second half of the nineteenth century. The Metropolitan Commons Act 1866 prohibited the inclosure of any common land within the Metropolitan Police District, and the Commons Act 1876 introduced a requirement for "benefit to the neighbourhood" to be demonstrated when applying to the inclosure commissioners for a provisional inclosure order. The commissioners were to insert conditions to further the benefit to the neighbourhood, including the securing of free access to "particular points of view", preserving trees and historical monuments and providing for recreation to be permitted on parts of the common at such times and in such manner "as may be considered suitable".[15]

The 1876 Act also made provision for bye-laws to be made by the owner, thereby facilitating the making of regulation schemes for the future management of common land.[16] Although the introduction of neighbourhood benefit as a criterion for inclosure marked the first recognition of community (or public) values in the way common land was to be used in the future, the 1876 Act was, on the whole, unsuccessful. In particular, the establishment of regulation schemes required the assent of the lord of the manor and of two-thirds in value of the commoners, a requirement that Lord Eversley considered "proved fatal to any general adoption of the Act".[17] Although the legislation was moving towards recognising community values in the urban commons, therefore, wherever there were tensions between property rights and community values, the former continued to have primary relevance to determine how the land should be managed for the future.

The Commons Act 1899 completed the shift in public policy to promoting regulation – as opposed to facilitating the inclosure – of the commons by introducing a simplified scheme allowing district councils to adopt and implement schemes for the management of common lands. When framing the terms of a management scheme, they were to consider including conditions to ensure benefit to the neighbourhood.[18] A large number of schemes were approved under Part 1 of the 1899 Act[19] and most included the provision of free public access to members of the neighbourhood of the common in question. Nevertheless, the 1899 Act conferred a veto on the making of a regulation scheme on the lord of the manor and/or one-third in value of the interests of commoners affected by the scheme.[20] Despite the rebalancing of the private and collective (or public) interests in the preservation, use and management of common land that is evidenced in the 1876 and 1899 Acts, therefore the private interest represented by rights of property retained a powerful influence over the future of the commons. Indeed, Lord Eversley considered that many more regulation schemes would have been applied for were it not for the lord of the manor's veto on their adoption and implementation.[21] Nevertheless, more than 61,000 acres of common was brought under regulation by schemes under the

1876 and 1899 Acts, and inclosure was at an end by the close of the nineteenth century.[22] The final step in the journey of public policy towards recognising public recreational rights over urban common lands came in 1925, with the enactment of section 193 of the Law of Property Act 1925. This remains in force today and gives all members of the public "rights of access for air and exercise" to all urban and metropolitan commons – a right that can be exercised on foot or horseback, but which does not give a right to enter land with vehicles or to camp or light fires on the land.

The modern legal context for recreational public access was completed by the provision for the registration of common land and common rights by the Commons Registration Act 1965 and by the introduction of the "right to roam" over large areas of open land by Part 1 of the Countryside and Rights of Way Act 2000. The 1965 Act was intended to preserve open spaces and protect them from commercial development.[23] It provided for the registration of two categories of land: common land and Town and Village Greens. Common land is not "community land", in the sense of being communally owned. Most common land in England and Wales is privately owned, but it is characterised as "common" by the existence of private property rights over it that give individual members of the community access to its resources. Alternatively, it may be the waste land of a manor, over which no common rights subsist.[24] Much of the modern law of commons is still premised upon the rights of commoners (appropriators) to take or use the resources of the land, as described above. It follows that English Law on commons governance offers a means for large numbers of people to share in the benefits of land that they do not own, but it is poorly adapted for the implementation of modern public policy.

The contemporary law governing public access to commons was completed by the Countryside and Rights of Way Act 2000, which introduced a "right to roam" over many categories of private land, but on terms that are closely regulated under the terms of the Act. Once it is mapped, common land becomes "access land" for open-air recreation by the public.[25] The land will therefore acquire a "public" character, under which a right of access is granted for any person to enter and remain on the land for the purpose of open-air recreation, subject to specific statutory restrictions.[26] These exclude the use of vehicles; horse riding, cycling or camping; the intentional or reckless taking, injuring or disturbance of any animal bird or fish; the intentional or reckless taking, damage or destruction of eggs; and any activities connected with hunting, shooting, fishing, trapping or snaring, taking or destroying of animals, birds or fish.[27] Some of these activities (e.g. fishing) may be permissible as a matter of private right with a *profit a prendre* granted by the landowner. In general terms, however, the public right of access under the 2000 Act facilitates and protects the right to open-air recreation over all common land, including walking, sightseeing, bird watching, climbing and running. Dog walking by members of the public is permitted over access land, but not professional dog walking. The landowner can also voluntarily dedicate a common for public access under the 2000 Act,[28] and this can be used to extend the right of access to people other than walkers (e.g. horse riders) and for activities not otherwise allowed by the 2000 Act.

The 2000 Act, therefore, created a set of general access rights subject to standard access conditions set out in the legislation itself, and this will be applicable to all common land that is not regulated by specific legislation applicable to the common in question. Its importance is mainly – but not wholly – in relation to rural and peri-urban commons, as most of the larger metropolitan and urban commons – including our four case studies described in Chapters 2 to 5 – are subject to specific legislation with specially crafted access conditions applied under a management scheme specific to that particular common. These are explored in more detail below.

The "Managed" Urban Common

Three of our case studies – Town Moor, Mousehold Heath and Clifton and Durdham Downs – are urban commons that benefit from specific statutory rights of public recreational access. In the case of Mousehold Heath and Clifton Downs, these were granted by nineteenth-century legislation that closely reflects the priorities of the Victorian "parks movement". The terms on which public access was guaranteed differ slightly in each case:

- In the case of Mousehold Heath, for example, the Scheme enacted by the 1884 Act requires Norwich City Council "to hold the Heath for the benefit of the inhabitants of Pockthorpe and the citizens of the city of Norwich with a view to their lawful recreation".[29] The Heath was to be regarded as a place of public resort and the conservators were to prevent (with some closely defined exceptions) encroachment or enclosure of any part of the common. Portions of the Heath could be set aside for cricket "and other games", in which case enclosure of the area concerned was permitted to prevent cattle straying onto the sports ground.[30] The fact that cricket was singled out for specific mention speaks to the middle-class perspective of the "ordered" recreation envisaged by the framers of the Scheme.
- In the case of Clifton and Durdham Downs, the 1861 Act provided for the common to "for ever thereafter be kept open and unenclosed as a Place of public Resort for the Citizens and Inhabitants of Bristol", but always subject to the commonable rights of the freehold owners of the Manor of Clifton and of others.[31] This latter qualification preserved (and continues to preserve) the common rights of graziers, those over Durdham having been subsequently registered under the Commons Registration Act 1965.

In both cases, the legislation was accompanied by the introduction of bye-laws which strongly reflect the Victorian approach to "managed" or "orderly" public access. These designate the common as a space for recreational activities, but expressly exclude a range of "communal" activities that may have persisted in the immediate past such as carpet beating, digging soil or taking vegetation for animal bedding or food. The Mousehold Heath scheme, for example, expressly empowered the conservators to introduce bye-laws to prohibit the cutting, digging, taking or selling of sand, soil, gravel, sods, turf "or other substances"; and also to prohibit the cutting, felling, firing or injuring of gorse, timber, heather, shrubs or other plants.[32] It also empowered them to provide for the removal and exclusion of various categories of individual deemed undesirable, including "card sharpers", "gamblers", "gipsies", squatters, vagrants, persons "guilty of brawling fighting or quarrelling" and any "idle and disorderly persons". Interestingly, the terms of the scheme implemented by the 1884 Act not only strongly reflect the imperatives notions of "orderly" recreation promoted by the Victorian parks movement, but also expressly reversed several rights previously enjoyed as common property rights by members of the local community, such as estovers (taking wood, gorse etc.) and turbary (taking soil or peat).

Notions of Belonging, Ownership and Identity

The oral history research carried out in the project sheds important new light on the extent to which prevailing governance and property systems in urban commons reflect notions of belonging, ownership and identity. This enables us to examine the extent to which "formal" legal and property classifications used in the governance of the commons reflect contemporary notions of ownership, use and the future community development of green space. It will also inform the conclusions in Chapter 10 positing a new multifaceted definition of the modern "urban common".

Long usage is associated in the public mind with concepts of "ownership" and "property". This can give rise to strong communal senses of identity, with the common as a key ingredient of "place" identity. This idea is reflected in the legal concept of prescription – the rule that long usage of rights over land can result in them being recognised as legal entitlements. Prescription at common law required that a right be used from "time immemorial" if it was to be recognised as a legal entitlement, and in practice use for a period exceeding 20 years would suffice.[33] But it is not reflected in the modern law governing the acquisition of rights of common – the Commons Act 2006 has abolished prescription as a basis for acquiring new common rights in the future.[34] And the decision of the court of appeal in 2016 in *R (on the application of Littlejohns) v Devon CC*[35] has meant that no new right of common by prescription can be claimed over land that was registered as common land under the Commons Registration Act 1965. The policy reflected in the 2006 legislation is to promote certainty of legal titles to land, but it also arguably promotes the interest of the "formal" freehold owner over the rights of members of the community to claim a public interest in the land based on long usage.[36]

This is, therefore, a question where contemporary formal notions of ownership and entitlement, represented in legal and normative property rules, diverge from those of many of the commoners that use our urban commons. Interestingly, the association between notions of community "ownership" and the long-standing exercise of common rights is especially strong where the rights in question have been exercised over several generations. The longer the period of community use, the stronger the link between the community's exercise of rights over the land and the generation of a sense of "ownership" and communal identity. This can be illustrated with some observations from the oral history participants in relation to Mousehold Heath (Norwich) and Town Moor (Newcastle):

> Mousehold Heath is part of the identity of the city … And it's a place where there's been change. I mean, you know, it's a historic part of the city as well.… It's deep in the history and identity of the city. And you know [the Victorian legislation] turned a largely sort of, slightly anarchic environment, at least from the point of view of, you know, the authorities, the cathedral and the city, the city council, into the kind of Victorian vision of a park.[37]
>
> It wouldn't bother me to go and [pick berries and fruit]. In fact, I probably do when I go with [my daughter], we probably do pick fruit, when we see it, we probably do pick blackberries and things like that. Because it wouldn't even occur to me that I wouldn't be allowed to … because I do feel ownership over it. I do……[but] if it was someone's garden, I wouldn't do it.[38]
>
> I think we look at it as common land, it belongs to everybody, basically. …..It is ours.[39]
>
> But yeah, people do feel a big ownership of it, and they think it's an unchanged place that they've known for many years.[40]

> Obviously the [Newcastle] Town Moor was part of our local heritage as well. And God knows what's underneath the soil. We were always told from, you know, a young age that the reason why it hasn't been built on is because it was protected by the Queen and that … whether or not that's a myth I'm not sure…..[and] in terms of my interaction with the Moor, … my whole life I used the Moor, not just for wildlife, but it's mindfulness as well.[41]
>
> People have rights to do things on it [Newcastle Town Moor] … But nobody owns it. And that's the beauty of it. I think the City Council holds it in trust as it were. For the city…..It is common land!.... The Freeman protect it because they are protecting their inherited rights and … to put their cows on it. But there's also a sort of subconscious awareness that the Moor belongs to everybody. I mean, are the other urban commons, similar to Newcastle's in any way? Is there an Act of Parliament protecting any other common land in the whole of the country? I bet there isn't …[42]

If we apply Lefebvre's theory to define the understandings of "space" revealed in these community responses[43] and to contrast them with the formal legal status attributed to each urban common, there is a clear divergence between our urban commons as "conceived" and as "perceived" space – between the understanding of the lawyers and planners and that of the public and commons' users. And, furthermore, the lived experience of many commons' users and user groups also differs. In other words, our urban commons display the different characteristics of perceived, conceived and lived green "spaces" simultaneously – their identity as community "space" is not only defined by their historical origins and legal classification, but also by the strong associations of "ownership" and belonging created by long public usage, and by their contemporary use as multifunctional open spaces for recreation and a multitude of other communal pastimes. Perhaps ironically, the divergence between (i) the formal legal understanding of a common based on ideas of prescription and the acquisition of common rights over the land and (ii) the perceived understanding of the community as to the nature of the land as "common" based on long usage, is getting wider. The abolition of prescription as a formal method of acquiring common rights over land makes the contrast between conceived and perceived notions of the urban common as public space more marked.

A New Approach to the Governance of Urban Commons?

As seen from the account above, the complex series of property rights and legal interests that currently subsist in our urban commons are largely derivative of and determined by the land use priorities of a past age. Many of our urban commons originated, as we have seen, in the medieval manorial system of land use when communal use rights were largely agricultural in nature. Many contemporary urban commons were formerly agricultural or rural, having become "urban" commons as large urban conurbations grew up around them. These origins are reflected in the common rights that still subsist over many urban commons, such as estovers, grazing rights and turbary. These rights now sit alongside statutory rights of public access created by private Acts of Parliament or by the Countryside and Rights of Way 2000. Balancing these interests and use rights can be difficult, and there is no overall guidance on how this should be done. It is done on a case-by-case basis by those with a vested interest in each urban common.

A new model for governing the modern urban commons should reflect contemporary priorities and public policy. A different approach could involve looking primarily not at land use rights (or "property" rights) in the commons, but rather focusing instead on the *benefits* that our urban commons provide for our communities. We should instead ask ourselves: what *ecosystem services* do our urban commons provide or *what ecosystem services could they provide* with appropriate management? Rather than focussing on who has *rights* over the common, we should identify the various benefits that an urban common can provide and develop strategies for maximising those benefits for our communities. A new approach focussed to identifying ecosystem services would also reflect modern approaches to valuing nature and reconnecting people and the natural environment.

This approach would focus on identifying the benefits – "ecosystem services" – to society at large and to future generations that our urban commons provide.[44] Ecosystem services can take different forms and are usually grouped under four headings[45]:

- *Provisioning services* include the products we obtain from natural ecosystems, for example, food, including meat produced using common grazings such as that on Town Moor in Newcastle.
- *Regulating services* include the benefits we derive from ecosystem processes, such as water purification and climate regulation. The ability of drainage systems on urban commons to mitigate flood risk or to purify water supplies for drinking water supply after treatment are examples here.

- *Cultural services* include non-material benefits we derive from our interaction with the natural environment. These include the preservation of precious landscapes, the promotion of biodiversity on our urban commons and the provision of open-air recreation. In the context of the contemporary urban commons, cultural ecosystem services of this kind are especially important, and in many cases have replaced provisioning services (e.g. the production of food or cutting of turbary or estovers) as perhaps the primary contribution that the commons provide for the community.
- Finally, *supporting services* are necessary to produce all the other ecosystem services from which we benefit. Examples here would include soil formation through suitable land management and natural nutrient cycling (such as soil breaking down animal or vegetative waste).

Developing a new approach to the governance and management of our urban commons with a focus on ecosystem services would require us to appreciate – and to navigate – the interactions between those ecosystem services that are captured in property rights, those that rely upon statutory rights (e.g. public access for recreation) and others that are not currently recognised in the normative ordering of the rules governing the management and use of urban commons. An approach focused to ecosystem services would also bring the management of our urban commons more closely into alignment with developments in modern public policy.

Developing a modern approach to identifying and managing the ecosystem services that an urban common can provide can be facilitated if we adopt a more expansive understanding of "property" notions. It is useful in this context to view property as a "web of interests".[46] The liberal common law understanding of property is to view it as a bundle of "rights" and categorise these as entitlements – to take produce, to transfer the land or other legal interest held, etc. But, as has already been noted, property rights also represent elements of resource utility, and these may allocate different use rights to different people simultaneously.[47] But resource utility in the case of an urban common will include not only extractive rights (the traditional concern of common law entitlements such as estovers) but also recreation, public access and biodiversity. Resource utility encompasses the provision of ecosystem services, in other words. These wider interests currently have diverse origins and definition in recognised property rights, in statute or in customary community practice; but viewing all these diverse elements as part of a single interconnected "web of interests" is a useful way to explain and understand the interactions between them and to organise their management for the future.

It is also important to recognise the impact that the classification of user rights as "property" rights can have on the balance of ecosystem services provided by a common. For example, many urban commons provide a rich and valuable biodiversity, and this can in some cases be a defining characteristic for their future development and management. In this context, property rights may be important in defining the historical context for the value of the biodiversity of each common. Of our case studies, for example, Town Moor in Newcastle was considered by participants in our oral history research to provide an invaluable resource for recreational activities, ranging from walking and running to public festivals such as Pride Newcastle (held annually in recent years) or the annual Hoppings Fair. But its biodiversity value is determined by its primarily agricultural nature as grazing for cattle and is poorer than might otherwise be the case. Mousehold Heath in Norwich, on the other hand, contains important remnant heathland habitats, and is subject to an active strategic management plan to improve both its value as a heathland habitat and its overall biodiversity.[48] But because the Heath ceased to be used as a common grazing resource following its purchase by Norwich Corporation in 1861, it no longer has a habitat conditioned by agricultural exploitation. This means that its biodiversity is not conditioned by the impact of livestock grazing, and the lack of registered common rights over the land (e.g. to graze, to take estovers or turbary) means that a strategic plan for biodiversity improvement can be implemented without the restrictions that might otherwise apply.

If we look back, therefore, the historical capture of the community benefit from commons in formal property rights may often explain their contemporary governance, recreational and biodiversity values. Looking forward, this can also have relevance to the development of strategies for their future governance and use. In particular, property rights retain some importance, in that they determine who can "sell" ecosystem services derived from or provided by the common.[49] Environmental policy, in particular, is now modelled on paying land managers for providing ecosystem services ("public money for public goods"). The UK government's 2018 *Health and Harmony* policy statement[50] promised a new environmental land management scheme (hereafter ELMs) to deliver the outcomes of the UK government's 25-Year Environment Plan and its Clean Growth Strategy.[51] This is being introduced from 2022 to replace payments formerly made to farmers and land managers under the EUs common agricultural policy. The central thrust of ELMs is to pay land managers for improving biodiversity, creating and managing wildlife habitats and undertaking environmentally beneficial land management actions. The ELMs is not restricted to agricultural land use, however, and might afford funding for landscape and nature recovery projects on common land. At the same time, many commons host important protected areas for wildlife such as Sites of Special Scientific Interest (SSSIs), Special Areas of Conservation and/or Special Protection Areas. Clifton and Durdham Downs, for example, host several SSSIs and Mousehold Heath hosts the St James' Pit geological SSSI. In addition to funding through the ELMs, funding can be provided by Natural England for the management and improvement of habitats within designated SSSIs.[52] But a management agreement for environmental management with payments can only be concluded with someone with a legal "interest" in the land concerned.

Public funding schemes to support environmental management in sites such as SSSIs or under ELMs therefore require the recipient of public funding to hold a property right in the land concerned. Funding is usually made pursuant to a land management agreement with a public body (such as Natural England), in which the property rights of the owner/land manager are exchanged or modified in return for payments for the duration of the contract. This explains, for example, why Natural England has a management agreement under the Countryside Stewardship programme with Norwich City Council – the freehold owners of Mousehold Heath – for the regeneration of its heathland habitats.[53] In the case of Town Moor, on the other hand, the Freemen of Newcastle hold herbage rights that could in principle be traded in an agreement for public funding to further develop the environmentally beneficial management of the moorland habitat on the Town Moor. The City Council as freeholders could also enter into management agreements with a public body such as Natural England for the management of the Moor, but this could not interfere with the herbage rights of the Freemen without their concurrence and consent. Although property rights are important, therefore, they cannot override or restrict statutorily conferred rights over an urban common; for example, rights of free and open public access for recreation conferred by the private Act of Parliament (such as the Town Moor Act 1988 or the Mousehold Heath Scheme Confirmation Act 1883) or by the Countryside and Rights of Way Act 2000. Neither can a management agreement between a public body such as Natural England and the holder of property rights in the common, for example, graziers.

Developing an Ecosystem Services Approach to Commons Governance

A new approach to the governance of our urban commons could, therefore, move away from the organisation of commons governance around property rights and statutory entitlements and take a broader view that identifies and promotes the ecosystem services that a common can provide. This would deliver several important benefits:

- It would give us an enhanced ability to *identify* the benefits that urban commons provide, some of which may be invisible to the community at large. Drainage is a case in point. Many urban commons provide this important regulating ecosystem service, one that is not immediately apparent. Town Moor in Newcastle, for example, provides a drainage service function that protects much of the urban centre of the city from flood risk – a factor of which most of the interviewees in our research sample were unaware. The urban commons also provide important cultural ecosystem services: access to recreational green space, opportunities to reconnect people with nature and the preservation of precious landscapes, all of which can have important physical and mental health benefits.
- It would also enable us to *balance* the different ecosystem functions provided by our urban commons more effectively. As we have seen, some of the ecosystem functions provided by urban commons are reflected in, and are captured by, property rights. But not all of them. Provisioning services, such as grazing livestock or taking water for irrigation or drinking, may be captured in property or riparian rights. But most cultural ecosystem services are not. It follows that there is no readily available mechanism for balancing these interests where their exercise gives rise to conflict.

Viewing these as part of a wider "web of interests" would be a useful starting point for the development of a strategic plan for each urban common, targeted at the maximisation of the ecosystem functions and services that the common can provide, both for the community and for those individuals with property entitlements in it.

Notes

1 Alfred F Beckett Ltd. v Lyons (1967) Ch 449.
2 Alfred F Beckett Ltd. v Lyons (1967) Ch 449.
3 Kevin Gray and Susan Francis Gray, 'Private Property and Public Propriety', in *Property and the Constitution*, ed. Janet. McLean (London: Bloomsbury Publishing, 1999), pp. 11–39.
4 See Commons Act 2006 s15.
5 See Alfred F Beckett Ltd. v Lyons (1967) Ch 449.
6 Countryside and Rights of Way Act 2000, ss 1, 2 and Sched 1.
7 See generally, *Gadsden and Cousins on Commons and Greens* (3rd edition by Edward Cousins, Richard Honey and Hugh Craddock) (London: Sweet & Maxwell, 2020) at 2–34 to 2–49; C. Rodgers, E. Straughton, A. Winchester and M. Pieraccini, *Contested Common Land: Environmental Governance Past and Present* (London: Earthscan, 2010), pp. 4–7.
8 See Chapter 2 p. 39.
9 The Mayor and Corporation of Norwich v Brown and others (1883) 48 Law Times 725.
10 See Chapter 3 pp. 50ff.
11 See the Report of the Royal Commission on Common Land 1955–58 CMND 462 at p. 162.
12 See Hazel Conway, *People's Parks* (Cambridge University Press 1991) at 23, 24. Conway cites the example of Nottingham, which was surrounded by commons to which the rapidly increasing urban population turned for recreation, including cricket. The recreational importance of the Nottingham commons did not prevent their enclosure in 1845 (see further ibid. at p. 24).
13 Inclosure Act 1845, s.12.
14 Inclosure Act 1845, s. 14. This provision also prohibited the enclosure of common land within 3½ miles of a town with 70,000 inhabitants, or within four miles of a town with 100,000 inhabitants. The population of each town/city was to be ascertained from the last relevant parliamentary census.
15 Commons Act 1876, s.7.
16 Commons Act 1876, ss. 15–17.
17 Lord Eversley, *Commons Forests and Footpaths* (London: Cassell & Co, 1910), p. 251.
18 Commons Act 1899, s.1.
19 Seventy schemes were established between 1899 and 1910; see Eversley, op.cit., at p. 259.
20 Commons Act 1899 s.2(4).
21 See Eversley, op.cit., at p. 259.
22 See Tables 1 and 2 in *Gadsden and Cousins on Commons and Greens* (3rd ed., London: Sweet & Maxwell, 2020), para. 1–12.
23 Royal Commission on Common Land, 1955–1958 Cmnd 462.
24 Commons Registration Act 1965, s. 22.
25 2000 Act, s. 1(1), Commons Act 2006 Sched. 5, para 7.
26 2000 Act, s. 2(1).
27 see 2000 Act, Schedule 2.
28 2000 Act, s. 16.
29 Mousehold Heath Scheme Confirmation Act 1884, Schedule para 3.
30 Ibid. Schedule, Scheme para 14.
31 Clifton and Durdham Downs (Bristol) Act 1861, s. 1.
32 Mousehold Heath Scheme Confirmation Act 1884, Schedule, Scheme, para 17.
33 The legal fiction was that the right was exercised in 1189 (on the accession of King Richard I) but that the document granting the right had been lost in the intervening period. Prescription could also be claimed under the Prescription Act 1832.

34 Commons Act 2006, s.6. This restriction applies to land within the "pioneer" registration areas to which the 2006 Act applies; this will be extended in the future to cover all registered common land.
35 [2016] EWCA Civ 446.
36 See Christopher Rodgers, "Nourishing and Protecting our urban green space in a Post-Pandemic World" (2020), 21 Env L Rev 1–5.
37 Oral History Interview Norwich 8.
38 Oral History Interview Norwich 9.
39 Oral History Interview Norwich 10.
40 Oral History Interview Norwich 4.
41 Oral History Interview Newcastle 11.
42 Oral History Interview Newcastle 13.
43 See Chapter 1 at p.13.
44 "The products of natural systems from which people derive benefits, including goods and services, some of which can be valued economically, and others which have a non-economic value". See *The Natural Choice: securing the value of nature*, Cm 8082, June 2011, at page 12. Available at: http://www.defra.gov.uk/environment/natural/whitepaper/.
45 See the categorisation by the *Millennium Ecosystem Assessment*. Available at: http://www.maweb.org/en/index.aspx
46 See C.A. Arnold "The Reconstitution of Property: Property as a Web of Interests" (2002) 26. *Harvard Environmental Law Review* 281.
47 See Kevin Gray and Susan Francis Gray, 'The Idea of Property in Land', in Susan Bright and John Dewar (eds) *Land Law: Themes and Perspectives* (Oxford, 1998) 15 at 39ff.
48 See Chapter 3 at p. 72.
49 See further Colin T. Reid and Walters Nsoh, *The Privatisation of Biodiversity* (Cheltenham: Edward Elgar, 2016), pp. 98–104; Colin T. Reid and Walters Nsoh, 'Privatisation of Biodiversity: Who Can Sell Ecosystem Services?' *Environmental Law and Management* 15 (2013): 12.
50 *Health and Harmony: the future for food, farming and the environment in a Green Brexit – policy statement*, 2018, p. 36. Available at: https://assets.publishing.service.gov.uk/government/uploads/system/uploads/attachment_data/file/684003/future-farming-environment-consult-document.pdf
51 Defra (2018) *A Green Future: Our 25 Year Plan to Improve the Environment*. Available at: https://assets.publishing.service.gov.uk/government/uploads/system/uploads/attachment_data/file/693158/25-year-environment-plan.pdf (Accessed 2 February 2022).
52 See S.15 Countryside Act 1968; S.7 Natural Environment and Rural Communities Act 2007.
53 See Ch 3 at p.73.

9 Present Voices, Future Directions

Introduction

In early 2021, as the COVID-19 pandemic took hold, community relationships with their urban open green spaces generally seemed, at first, as if they might collapse. Moving quickly through phases of highly regulated separation from those spaces to quasi-illicit rendezvous with and in them, to tentative re-acquaintance, involving new yet uneasy relationships of use and sometimes abuse, there developed instead a refreshed understanding of community relationships of mutual dependency with those same green spaces and with their urban commons in particular.[1] Both this co-reliance as well as the relative lawlessness of the urban common were brought into sharp relief during the early stages of the pandemic. As the various waves of lockdown took hold, it became clear that these spaces were now crucial to those who had no outdoor space at home, who needed to care for and entertain children or who had to exercise outside of the now closed gyms and pools.

It is important to remember that not all responses to lockdown were negative. One person, local to the Town Moor responding to a student intervention probing health and well-being on the Town Moor, commented, "it was good actually. Was able to enjoy quality family time, and I also began picking up different skills and activities as hobbies", and another, "I really enjoyed the lockdowns, I was alone but not lonely. I was able to work on myself, read some books and get into a better routine".[2] Aleksandr Derkac, the young adult participant from Sprowston Youth Engagement Project, spoke about how the lockdown gave him time to start a fitness regime jogging on the heath:

> I used to go on jogs and walking and running because it let me have that time of peacefulness and mindfulness, and also allowed me to let all my emotions out and I just felt really nice and peaceful after it.[3]

In general, however, experiences of the pandemic were negative. Either way, increased free time and working from home, coupled with government-sponsored messages of fitness and anti-obesity, ensured that common land was and continues to be used for community recreation, just as it has for hundreds of years. The sharpest effects of the pandemic have passed, but there is a palpable legacy of

DOI: 10.4324/9781003204558-12

concern for individual and community access to urban green open space with initiatives as diverse as 15-minute cities and various open-space related "levelling-up" initiatives.[4]

Anthropologist and economic geographer David Harvey argued that "the right to the city is far more than the individual liberty to access urban resources: it is a right to change ourselves by changing the city",[5] and the urban common is a key "urban resource" in this important sense; an arena for the playing out, sometimes uncomfortably, of the urban imaginary; a city's hopes and desires, its dreams and sometimes its nightmares. In this context, can we see quasi-resistive and certainly contested acts of barbecuing and Frisbee-throwing alongside brief nitrous oxide highs with its resultant litter-dropping, (just some of the offences cited on our urban commons during lockdowns) as a means of escaping or at least releasing pressure from the destructive effects of school closures, isolation, domestic pressures and the general rolling back of hard-won, if still very partial, gender equalities?[6]

Through a range of community engagement projects and oral histories (conducted at first virtually and then in person as COVID restrictions relaxed), this chapter examines how contemporary commons' voices talk about, dream about and behave in their urban commons; the convergences and sometimes misalignments between their imaginary and their physical urban commons and how new interrelations between the tangible and intangible commons may chart a course towards a different future.

Norfolk-based geological specialist and biodiversity and environmental tutor Tim Holt-Wilson put this interrelation between the artificial and the natural, the disciplinary against the undisciplinary and conceptual very clearly:

> ...going back to the theme of encountering the physical record of the earth's changes, I think of geology as environmental history. I think that's the way to communicate it to people. Not to say 'geology', I mean geology is too important to be left to the geologists. It has to be communicated to people in a way that enables them to value the earth. And when I say the earth, I'm talking about everything, the biological and the geological together. I see the earth as a complete phenomenon, and my attention focuses on geological facts, it also focuses on aesthetics as well. When I see a geological exposure, I see something, I experience it aesthetically at the same time as I experience it scientifically.[7]

The project team's focus on four distinctive urban commons is grounded in the idea of the commons as a physical site. However, contemporary thinking about the commons has shifted attention towards the social practices or everyday actions through which a common is produced and maintained. By inviting city dwellers to participate in creative co-inquiries with artists and researchers, the project team planned not only to capture how people inhabit their commons with actions and imaginings but to re-imagine them as a process of commoning[8] – enabling, connecting and networking different forms of knowledge. We also hoped to understand something of the attitudes of the various urban commons' stakeholders towards the future of their open green spaces.

In addition to the oral histories described elsewhere, therefore, the project team instigated and developed two broad programmes of engagement activities.[9] The first, called "Urban Common Futures", included a number of community participation projects: *In-Common Sites, In-Common Walking, In-Common Sites In Action* and *Writing the Commons* described below. In parallel, a second group of projects, called "Learning through the Commons", involved a number of architecture and design student-led projects which became catalysts for a radical rethinking about dwelling on urban commons. "Wastes and Strays", as a whole, concluded with an end-of-project travelling exhibition we called "Wastes and Strays ....on the Road". Much of the textual narrative of this chapter is developed from the "In-Common Walking" conversations on Mousehold Heath. Responses and analysis from Town Moor, Valley Gardens and Clifton and Durdham Downs are collected as mostly visual work in the last section of this chapter as well as being outlined and analysed in case study Chapters 2, 3, 4 and 5.

Urban Common Futures

In-Common Sites

The project team used participatory arts, performance and design-based practices to collate and understand a range of cultural interpretations of urban commons and to gain a better understanding of how a creative co-inquiry approach engages different communities in research as co-producers of knowledge, and took a participatory action research (PAR) approach.[10] Community participants engaged in a series of self-directed, creative fieldwork tasks that enabled them to explore their urban commons in their own time while maintaining current social distancing rules.[11] These creative inquiries invited participants to walk, observe and interact with their urban common field sites using creative tools such as map-making, storytelling, sketching, the re-enactment of historical and everyday practices, creative writing, reflection, photography and field-recording. Taking a thematic approach, we aimed to amplify the potential for participants to reflect on commonalities and differences across the four urban commons field sites and their lived experience. The activities generated 'collections' of interpretative responses, which were collated into a series of artworks for the 'Wastes and Strays ....On the Road' exhibition.

In-Common Walking

The sensory, ethnographic approach used in this study was a combination of walking, mapping and reminiscence, reflection and the use of audio and field recording to document everyday practices. A series of one-to-one, in-situ walking conversations with urban commons users were conducted. The conversations were designed and conducted by Wastes and

Strays project member Siobhan O'Neill and participants were encouraged to guide O'Neill on a walk of their choosing and engage in a conversation around their pastimes, lived experience of the urban common and ambitions for its future.

In-Common Sites in Action

'In-Common Sites in Action' was a performance-for-film activity, co-created by Siobhan O'Neill and filmmaker Annis Joslin in partnership with the Sprowston Youth Engagement Project and designed to explore young adults' lived and imagined encounters with Mousehold Heath. Using participatory performance and film-based practices it generated a cultural interpretation of the urban common to gain a better understanding of how an arts-based approach might engage young adults in research as co-producers of knowledge.

Like 'In-Common Sites', this study adopted a participatory action research (PAR) approach. It tested out a strategy which used a combination of methods including walking, sensory mapping, performance and filmmaking to document everyday practices, enabling participants to tap into and describe their commons knowledge and expertise. In the study, members of the Sprowston Youth Engagement Project participated in drama-based workshops facilitated by O'Neill. The participants collaboratively explored how young people use and value Mousehold Heath and how best to articulate their visions for urban green space in the future.

(See Figure 9.7)

Writing the Commons

Wastes and Strays project member John Clarke worked with the Avon Gorge and Downs Wildlife Project, and the Friends of the Downs and Avon Gorge as part of a short creative writing 'residency' on Clifton and Durdham Downs, Bristol in 2021/22. In the interval between lockdowns, he ran workshops with local schools and the public to explore how creative writing may combine historical research about urban commons with contemporary experiences of these important social and ecological spaces.

Learning Through the Commons

Running in parallel to the "Urban Common Futures" suite of participatory action research and co-creation activities were a number of architecture and design student-led projects centred on Brighton's Valley Gardens and Newcastle's Town Moor. These projects included one small structure or "pavilion" designed by students at the School of Architecture at the University of Brighton and another pavilion as well as a "participatory action research" series of projects by Newcastle

University students, alongside projects set on Brighton's Valley Gardens by University of Portsmouth architecture students.

Pavilions of the Commons with Students from the University of Brighton and Newcastle University

The project team developed a performance brief for a pair of small mobile structures designed to be catalysts for community engagement and as tools for thinking about urban commons local to the students involved. Architecture students in the School of Architecture, Technology and Engineering at the University of Brighton, led by their tutor Graham Perring, designed theirs for Valley Gardens, while design students in the school of Architecture Planning and Landscape at Newcastle University, led by Daniel Mallo and Armelle Tardiveau, designed theirs for the Town Moor. As a model of, and a mirror to, multidisciplinary collaborative design, these Wastes and Strays "pavilions of the commons" proved to be powerful apparatuses for thinking about co-operative design processes.[12]

(see Figures 9.2, 9.4 and 9.5).

Urban Commons: A Collaborative Project Exploring the Future of Urban Commons with Students from Newcastle University

> Also led by Daniel Mallow and Armelle Tardiveau in collaboration with Wastes and Strays project team member Siobhan O'Neill. The project, involving Newcastle students working with the community of Town Moor users, aimed to:

- Unpack social histories of the Moor as well as reveal traces and echoes from a distant or close past.
- Explore the Town Moor's current uses and users, including its legal status and governance, as well as foreground social stories and existing biodiversity
- Critically consider the future of commons and their pivotal role for community, health and ecology in the twenty-first century.
- These issues were explored through the articulation of design-based research methods relevant for engaging with communities during the time of pandemic. Through the design and trial of a series of engagement prompts, students were tasked with "probing" the past/present/future of the Town Moor.
- Phase 1: revealing histories and traces of the past.
- Method: signage project.
- Phase 2: unpacking stories and narratives of the present.
- Method: participatory research pack (cultural probes).
- Phase 3: trialling the future.
- Method: planning and making a spatial or performative intervention (signposting the future).
- Phase 4: portfolio narrative.
- Production of a Design portfolio.[13]

(See Figure 9.3)

Urban Commons: How to Build an Architectural Commons with Students from the University of Portsmouth

> Architecture students from the University of Portsmouth School of Architecture led by Wastes and Strays co-investigator Alessandro Zambelli and his teaching partner Yana Nanovska set students a two stage project:
>
> Stage 1. To design a temporary pavilion structure on Valley Gardens in Brighton as a "built manifesto" for students' reactions to researching this urban common. This brief was an adaptation of the "performance brief" given to the Brighton and Newcastle students.[14]
> Stage 2. To develop the "manifesto" pavilion into a larger urban strategy and eventually a building.

(See Figure 9.6)

Wastes and Strays ....on the Road

In May and June of 2022, as "Wastes and Strays" entered its final weeks, we took the work of all three work packages, "past", "present" and "future", on the road. A series of one day exhibitions at each of our four case study urban common locations sampled public reactions to and engagement with excerpts of the work of the entire project. This "travelling show" visited the Wylam Brewery, Newcastle (10 May), Clifton Observatory, Bristol (16 May), the Open Market, Brighton (2 June), concluding with a screening of the Sprowston Youth film at Frere Road Community Centre, Norwich (7 June). The exhibition showcased samples of the artworks, poetry and performances generated through our creative co-inquiries as well as overviews of the historical and legal research conducted by the team, all collected together to stimulate further exchange and conversation.

Work and Play on Mousehold Heath

If, as we have argued in Chapter 7, the urban common is a site of contestation, then one of the conflicts most clearly played out is that between the work and leisure practices of their past and present users. As far back as 1964, in Denman's seminal post-war assessment of common land generally *The Future Use of Common Land*,[15] the categories of concern seem very familiar and are echoed in conversations with the current stakeholders of our case study urban commons:

> (a) to make good the present neglect and prevent its recurrence; (b) to satisfy the local claims of landowners and commoners; (c) to meet in some measure the public demand for recreation; (d) to protect scenic beauty and preserve the habitats of natural flora and wild life; and (e) to weigh these private and public claims against the competing demands for land from housing, industry, agriculture and other essential pursuits[16]

Figure 9.1 "Wastes and Strays …On the Road". Publicity poster with the Brighton event highlighted.

Katrina Myrvang Brown identifies work in its industrialised form as one of the causes of commons enclosure, "the interrelated pressures of population growth, commercialisation, industrialisation, successive rounds of enclosure legislation, and an intellectual and cultural privileging of individual forms of property, have all contributed to the extinguishment and undermining of communal resource rights".[17] In more recent times, urban commons have relied on a different kind of labour to sustain them. Certainly, volunteer labour on Mousehold Heath is essential for its maintenance. The goodwill of a volunteer labour force is explicitly proposed by, for example, Colding et al. as key to their success: "many urban green commons hold potential to reduce management costs on public lands due to that they draw on volunteer-based management and by the self-interest of the participants".[18]

The volunteers on Mousehold Heath have a strong sense of duty as suggested here; duty to the land as well as a duty to others. Volunteer and Associate Professor in Public Policy and Public Management John Turnpenny describes his weekly litter-picking duties:

> …of course, there's more litter than anyone could ever pick up in a whole lifetime. But it would be very easy to go, 'Oh, well, there's no point'. You know? You pick one piece up, there's 20 bits you've missed. And then I thought, 'Well, actually, if I spent three hours picking up litter, I've actually done as much as I could in that three hours'. I've done my bit. This is my role. That's a bounded role, it's not just about me, it's about other people as well.[19]

Norwich Geography teacher Owain Hall picks up on this in relation to younger people:

> […] I have girls coming to me and asking if there are any litter pickers sticks that they can use to go and pick up litter around the school grounds, because they enjoy doing it. And I think that a shift that's definitely taken place.[20]

Some volunteers, like Paul Lynch, describe a strong emotional attachment to their work too: "If you're doing some work here, you are emotionally responsible for it. I feel emotionally responsible for it".[21] But as John Turnpenny himself concedes, "I think it's great to rely on volunteers, but volunteers can walk away if they don't like it".[22]

Over the centuries, the balance of work to leisure activities on the commons has shifted markedly to the latter. Jack Anderson argues that the "right to recreation" has long been contested on the commons. Using an account of legal proceedings from the mid-seventeenth century, he recounts that "the Court of King's Bench held for the villagers holding that the right in question was customary in nature and a 'good custom' at that because 'it is necessary for inhabitants to have their recreation'".[23] Arguing also that sport in particular has been seen as "civilising" might account for various landowners and political elites allowing such recreational events to happen on otherwise lawless commons. It might also help to explain why unregulated sport and recreation still creates unease; cycling, dog walking, jogging are all from time to time the subject of conflict on our case study urban commons.[24]

Nevertheless, these activities continue to grow in popularity and, in relation to the continued health of the elderly and the future health of the young, are seen as an important aspect of the future of urban commons.

In relation, for example, to the much-contested BMX biking in public green open space – an occasional flashpoint on Mousehold Heath – Katherine King and Andrew Church have observed that:

> existing research on lifestyle sports lacks consideration of young people's use of green spaces and the approaches of managers to conflicts in these spaces. Young people's experiences of leisure are closely tied to those who oversee their use of leisure spaces. [...] Findings reveal young people seek opportunities for autonomy in green spaces through mountain biking but contest normative management practices.[25]

BMX in many places in the world is seen as both an attractor of conflict but also a possible social adhesive. At a contested BMX track in Detroit, Morcillo Pallares (2016) observed that "the fragile maze of groomed trails, ramps and jumps require constant maintenance by the group that uses them and as a result strengthens the relationship between the community and the revived space".[26]

In our interactions with BMX bikers on Mousehold Heath, many people we spoke to were at pains to remain neutral or at least seem reasonable when asked about changes made to the landscape of the Heath by bikers. Some, though, were demonstratively supportive, as outlined in Chapter 3. Others, like Paul Lynch, were conciliatory:

> the only way to stop that is to get together with them and just talk it through, see if there can be a compromise. [...] it's not going to work if you start shouting at people to stop doing something; they will carry on doing it [...] It's working with them, trying to work out a compromise [...] Because in some ways the kids will be kids. [Laughter].[27]

There is even tension between activities which are themselves contested. Local Forest School organiser Nicola explains, "because... there's a conflict as well because some dog owners don't like bicycles, some of them do, they're fine, but some of them don't".[28] Dog walking and to a certain extent BMX biking happen in all weathers, but some leisure activities are necessarily seasonal; "this is the place to come sledging in Norwich", explains Nicola:

> and this is where we used to come for a quick ballgame and things when our son was younger, and now with my friends' children [...] it's good for watching fireworks as well. And people really like coming up here for the view.[29]

For Nicola, the mix of leisure activities is even seen as a benefit in its own right:

> a lot of people all over the city drive here as well to then use it. That's why there's all these car parks everywhere, and they drive with their dogs, and

> they drive up here, especially now for sledging and for the fireworks, although they didn't happen this year[30], but in previous years they have to close the road because it gets so busy.[31]

In a similar vein, Mousehold Heath volunteer Rosamund Chetleburgh speculates about the causes of the then recent increased use of the Heath:

> I think it's a year-round thing. It is definitely noticeable that there were quite a lot of children going past very politely on their bikes and making use of it. I think it is in continual use, but I wonder if maybe some of that was pushed by lockdown and there not being other things to do.[32]

Later, wondering what the future might hold for use of the Heath post-pandemic, Rosamund Chetleburgh speculates that:

> it would be lovely to use it as an event maybe, I don't know if they need to raise money for equipment or something, to have a May fair and then the proceeds at a charity event, like some part of whatever you take to buy more equipment that people, volunteers need.[33]

Living Well on Mousehold Heath

The need for the protection of existing and the creation of new urban open green space is ancient, but in recent years, this need has become pressing.[34] It became evident during the lifespan of "Wastes and Strays", especially in light of the COVID pandemic, an event unforeseeable at the time of the project's inception that proximity and rights of access to urban commons and common-like open space were vital for community cohesion and resilience as well as individual health and well-being. The World Health Organisation cites evidence for "improved mental health and cognitive function", "reduced cardiovascular morbidity", "reduced prevalence of type 2 diabetes" and even "improved pregnancy outcomes".[35] Most recently, a global "15 minute city" initiative has gained traction, as well as some controversy. The originators of this idea, C40 Cities, describe its aims as "compact and resource-efficient cities, with co-location of residences and jobs, mixed land use and good access to public transportation [which] includes community-scale education and healthcare, essential retail like grocery shops and pharmacies, parks for recreation, working spaces".[36]

That open green spaces in general are loci of health and well-being is well understood, but it also resonates with what we found speaking to local communities. In particular, people were keen to talk about the effects of the COVID-19 pandemic and to reminisce about the times of lockdown. John Turnpenny observed that:

> a lot of people were here a lot more during the lockdowns. I certainly noticed that [...] probably twice as many people than [...] now. [...] I think it was good that people felt they were able to come up here and use it, use the space.

And this, for him, seemed to bode well for the future, "maybe people discovered it and hadn't really been here before".[37]

Walking is perhaps the most popular and the most straightforward of health-related activities on the Heath, "solvitur ambulando"[38] as local historian and community worker Colin Howey put it, yet the Heath has been and continues to be used as relief for more debilitating conditions. Colin Howey describes an organisation called Pit Stop, where he volunteers, "working with men in recovery, with mental ill health". Where he describes "doing a series of history walks with them", reads to them:

> a little quote from George Borrow, the Victorian Norfolk writer born in Dereham [...] for me it kind of sums up a bit about the potential for stillness, serenity, slowing down your inner clock, and I think it's rather beautiful. I probably won't read it very well, but I think it's rather beautiful. 'There's night and day, brother, both sweet things; sun, moon and stars, brother, all sweet things; there's likewise a wind on the Heath. Life is very sweet, brother; who would wish to die?'

"My point in reading it to them wasn't simply that this happened here, this is history".[39] Colin Howey is keen to emphasise, "it's that there's something here for us to carry forward". Mousehold Heath local historian Trevor Nuthall in the same conversation sums up the advantages of these kinds of healing, "so, there's physical and spiritual. It's easy to underplay in our kind of materialistic age the benefit of the spiritual side of it".[40] Rosamund Chetleburgh believes this mode of being in the world is particularly important for the next generation of users of the Heath:

> there's a lot of young, youngsters, you know, 12, 13 slightly younger, bit older, but on Sunday morning they all had their bikes out there. And they weren't on a computer, they were doing things like cycle rides with mates and using the Heath. It's got to be a good thing for them.[41]

Tim Holt-Wilson also links mental health to healthy futures more generally:

> I do think that if you can link the outdoors to people's health outcomes [...] we know that there's a lot of need in parts of Norwich [...] we know that some people[s] [...] lives have gone very, very sour through lockdown [...]. I think that linking the outdoors and rewilding to mental health outcomes and wellbeing is a very, very positive thing that we can do, and we can piggyback all sorts of useful initiatives on the back of [it].[42]

Rewilding or just wilding has become a highly contested mode of increasing biodiversity in recent years.[43] Relatively young as a discrete strand of ecological thinking, rewilding as a form of biodiversification became academically respectable with the publishing of *Rewilding and Biodiversity: complementary goals for continental conservation* in 1998,[44] and more recently has taken a firm hold in the public imagination. In the UK, for example, Rewilding Britain is an active and thriving network and has spawned many more local offshoots together with sister organisations in Europe, the US and Australia.[45] Talking to us, retired National Health Service (NHS) administrator and conservation volunteer Robert Stubbs reminds that "the whole of the landscape, everywhere apart from primal forests, are one hundred percent man made […] So, everything has to be managed. Whereas a lot of people don't understand, you can't just let nature take its course".[46]

The tensions between heath and woodland, managed versus wilded and even more fundamentally natural against artificial are played out on Mousehold Heath and amongst its managers and users, even sometimes between their memory and the historical record: "you have bracken which is very […] enveloping of all these trees. Whereas you look at historical […] pictures of the area and it's completely clear. There aren't much trees. There isn't […] bracken at all".[47] It's not at all clear that true wilderness can even be recognised, as Trevor Nuthall warns, "we have this notion of it being wild, but it's always, up until relatively, well, even now with Will's work it's managed".[48] Paul Lynch too reminds us that there is a fine line between managed wilderness and any notional "true" wilderness, and that to maintain the distinction, there is real labour involved:

> well, here it's extending the heathland by taking out invasive birch, that's the main job today, and the bracken […] if you leave it, you get the birch moving in, but once you've got the birch moving in you've got the bigger trees like ash and oaks, in some areas, beech. So, the big trees will naturally take over, given time[49]

John Turnpenny on the same theme said, "you introduce one thing and that has all sorts of knock-on consequences. We're already seeing species that weren't here in the UK 20 years ago and they're becoming very common now".[50]

Colin Howey was keen to remind us that human intervention in the landscape is ancient and episodic:

> if you go back to the Domesday survey, we know that there were over 1,000 swine that were herded up here, which is going to need an extensive area, it's going to need a ditch and a bank probably with hawthorn growing on it. The pigs are going to root up all the groundcover so it's going to be quite bare and open on that part[51]

That areas which are wooded now may not have been so historically seems to raise the question of which period in history is most authentic, but, in fact, demands that commons stakeholders make informed decisions about what point in time is their preferred reference point for recreating any putative "past" landscape. Colin Howey's answer anticipates this notion of false authenticity, "you'd have to go […] not that long after the Ice Age if you want that".[52] Paul Lynch agrees: "there's no such thing as ancient woodland. It's always been heavily managed, very heavily managed. Especially around here".[53] More profoundly for Colin Howey, "human beings have been working with nature and you could argue we are part of nature as well. We are living organisms, it's just that we're not always good at finding harmony with it".[54]

Nicola describes the current small but for her significant battles for a biodiverse future, some of them being won and some not, "there's so many dogs now and everyone runs and going in the pond that last year there was so fewer frogs and frogspawn. It was very sad. But hopefully they will get their numbers back again".[55] The day-to-day trials of even very small-scale biodiversity projects include for her, "trying to educate people about dogs, because I think that's why Will thinks there aren't any snakes here anymore".[56] And on another occasion, she recalls:

> with one of the schools that I used to work with we made a wildlife garden at Plumstead Road, a butterfly garden, it was called Butterfly Garden, and half of it's become a car park now, unfortunately. But I'm in a battle [...] So, but I'm in, I haven't given up. [Laughter]. I'm going to battle out somehow, because I don't, yeah, so that's, yeah.[57]

Reflecting more generally, Nicola insisted that "everywhere we need to increase the biodiversity, because we're, [...] in an ecological emergency and a climate emergency. That's why we need the bicycles. Bicycle paths and footpaths and wildlife".[58] But this vision is multifaceted; even the sledging that Nicola talks about elsewhere, far from damaging the Heath actually has unintended beneficial consequences: roughing-up the surface of the grass on the hill stimulating the growth of meadow wildflowers.

Climate emergency and its consequences for Mousehold Heath is on the mind of John Turnpenny too, "I'm particularly concerned about [...] climate change and how that's going to affect our landscapes and what that's going to mean for heathland and the animals and plants that live here", he begins, "I don't think we have fully understood just how much of an impact that could have. [...] it's only when it starts to happen that you notice all of a sudden plants that were here don't exist here anymore". More speculatively, he wonders whether:

> there's an argument for saying there's a certain amount of climate change that's just going to happen anyway so we should try and adapt to it, and that's encouraging more plants that can deal with drought in the summer and floods in the winter.[59]

From Rosamund Chetleburgh, there is clarity about the proposed outcomes, about what this biodiverse vision for the future might look like, "I was once on a run and suddenly there was a deer coming one way and me coming the other and we went, 'Er'. It was so lovely to encounter a deer on a summer evening". And then, reflecting on the future:

> I really would like to think that the children, young people, that in the future there still will be like hedgehogs and foxes and green space and some remnant of the continuity of ages, if you want. It's a bit romantic, isn't it? But yes, it does matter to me to leave something for them.[60]

Mousehold Heath, as with our other case study urban commons and a multitude of other types of urban open green space, must be thought of as pieces of a jigsaw demanding to be connected with one another; the future of urban resilience lies in no small part in these necklaces or "corridors" of biodiversity. According to the Green Infrastructure "meso scale", a concept developed by Barker et al.,[61] the environmental benefits of "Green Corridors" are, of course, the "facilitation of species movement" and the associated increase in habitat area, but also include the reduction of urban heat islands and improvements in water and air quality. Peter Shirley has also emphasised a "deeper green view" which sustains "the biosphere of which human beings and their settlements are but a part".[62]

One threat to urban commons is their isolation – their remoteness from other public green open spaces, often as a result of road-building around their remaining perimeters. Proposals to unite spaces like this by connecting them into "green" corridors emerged in the 1980s, with a report published by the London Ecology Unit to the Greater London Council where "recognition [was] given to the need for green corridors extending through the city, linking areas of wildlife habitat and including other categories of open land, such as parks, allotments, river valleys, reservoirs or railwaysides". This connectivity or perceived lack of it, chimed with the concerns of a number of locals at our case study urban commons. Trevor Nuthall, speaking in relation to Mousehold Heath, also emphasised mobility, but for him this meant human physical movement as well as a kind of social mobility:

> to have this huge, great green lung, if you like, in an urban setting like this [is] a place where lots of different people […] cross […]. Sometimes meeting, mixing, mingling, oftentimes not. They occupy different social spaces. They come here for lots of different reasons and purposes.[63]

Speaking specifically about these issues at Mousehold Heath, Robert Stubbs warns, "[…] it's an isolated pocket. We've got no corridors from the wider landscape into this area […] You don't want to get isolated groups of wildlife. You need a diversity coming in".[64]

Where Exactly Are Our Urban Commons?

The past and current ownership structures of our four case study urban commons are complex and to the local people we spoke to – who weren't directly involved in the management or maintenance of the commons – either complicated or unknown or perhaps even seemingly unknowable. Some brief quotes from our oral history interviews illustrate this point:

Town Moor

"… in my head I would have thought it was Newcastle City Council [which owned Town Moor]. But I have never known, and I've never sought to know, because it's just been there."

"I would assume the council. I know there's like a […] I think there's a committee that kind of manages it. But I'm not sure who actually owns it."

"Ah, now that's interesting. It's very interesting because I believe it is owned by the […] or at least the guardians of it are the Freemen of Newcastle, which are this […] medieval group. So I think that they own it, but it's very complicated in Newcastle because I don't know if they hold it on behalf of the people of Newcastle"

"Well, it's, it seems all kind of in the mists of time, that it arose from kind of city guilds and crafts, craft's guilds and, but I don't really understand the complexity of that"

Clifton and Durdham Downs

"I think the Downs is a common ground, so it must be commonly [owned], but of course it has to have some form of ownership. So, I know that the Merchant Venturers and the city council are the people who maybe, probably, I imagine, co-own the Downs."

"Well, I guess 'Own' is a very specific question, isn't it? So, I believe that the people of Bristol have the freedom and rights to use the Downs […] Legally, my understanding is that […] the land is owned by a combination of the council and an organisation called the Downs Committee. And… I think there's an organisation called the Merchant Venturers in Bristol […] So, so yeah, I would suggest it's a combination of the council, the Downs Committee and the Merchant Venturers."

Mousehold Heath

"In my understanding, it's common land but that does not mean, as my understanding, that it is not owned by somebody. That […] I believe they can both be. So, if anybody owns, if anybody […] somebody must own it, I suppose. Norwich City Council probably own it in a sense […] I do find that really difficult."

"Well, the city does own it […] We are effectively just managing agents, the Conservators. So, I mean effectively […] the city does own it, and if they… I mean they may well feel they exercise enough control through the Conservators anyway"

"I think it's still common land. I don't know. I assume so. So, I think, yeah. Yeah, I don't even know […] I think you could ask anybody in Norwich, and they would say, 'Well, it's us, isn't it?'"

As Short and Winter point out, there has also, historically, been confusion about the difference between management and governance in matters to do with commons. For them, management is "physical and economic management activities and the day-to-day decision-making associated with these" and governance is "the decision-making structures, mechanisms and systems of administration which influence the operation of management systems".[65] For our urban common interviewees, this was a fine distinction. As John Turnpenny argues:

> I think the way open spaces are governed is really important. [...] that we have a warden who is an employee of the city council and has that direct relationship with the elected members and the officers. [...] the more kind of contracting that you have, the more it kind of gets disconnected from its original purpose, and that contracting out of public services is something that's really common.[66]

There is also a sense, from some, of a necessary watchfulness guarding against the erosion of rights. John Turnpenny describes his:

> awareness of how quickly something which is governed as a public space and for the common good can slip very much into private hands and private interests. Who benefits, who pays, who bears the cost? I think those kinds of questions need to be asked all the time about everything in public life, but about open spaces generally.[67]

As Trevor Nuthall argues, "lots of rights go because people [...] with those rights don't [...] bother anymore".[68] John Turnpenny returns to the idea that the key relationship is between those that tend the Heath and those elected to have power of governance over it, "I think good governance with close relationship between democratically elected people and the people on the frontline maintaining working is important".[69] But not everyone is for a completely bottom-up approach to the governance of the Heath, "I think it's important to have a wide range of input. It's also important that somebody takes the lead [...] trying to do things by consensus often ends up with nobody happy".[70] Others talked of having a sense of ownership over place, perhaps even in spite of the legal ownership structures of their commons. Rosamund Chetleburgh has a sense of this, saying, "a sense of belonging is hard to find, and it is incredibly good for your head". Pushed on whether she felt a "sense of belonging", she answered proudly, "yes, I do [...] my patch".[71]

We wondered whether locals and users of urban commons had a sense of their urban open green space being specifically common land or at least being common-like. Lindsay Campbell and Anne Wiesen, in the influential *Restorative Commons* and writing from a North American perspective, argue that "parks, community gardens, building exteriors, rights-of-way, botanical gardens, urban farms, vacant lots, public housing campuses, and closed landfills", are all examples of "fragments of the commons",[72] although it seems there may be some conflation here of common (or common-like) land with broader notions of "the commons". We found that there were perceived differences between commons and parks, "whereas parks tend to be very organised, Mousehold Heath is only organised in some kind of parts of it",[73] as Trevor Nutall put it. Or as Colin Howey, talking about the Heath remembers it, "it was a bit feral. Which is kind of, quite in-keeping with its history, in some ways, because it's always been an area that's sort of on the margins and stuff".[74] Being "on the margins" is, as we have seen elsewhere, a very apt description of urban commons.

Urban commons seem to attract a certain kind of lawlessness. John Turnpenny, volunteer conservationist, speculates about what he feels might be the attitude of others in this respect:

> …this is free. I want to use it and I'm going to use it whenever I want and for whatever I want. If I want to light a barbeque, I'm going to jolly well light a barbeque. It's my land. I pay my council tax; I'm going to do it'. And people might not […] invoke the idea that it is common, a commons, or anything like that.[75]

Clearly echoing Hardin's notion of the tragedy of the commons,[76] though he is doubtful whether others see a difference between a common and other places, "whether people see it as […] special and different enough to give it its own particular consideration, I don't know. Is it a destination place? Would people actually come here from quite a long way around? I don't know".[77] And later:

> we're all different, we all like different things. Some people like loads and loads of trees, some people like formal gardens with kind of pansies 20 centimetres apart and nothing else. Well, fine. If it's common, if it's common, it's got to be common, surely. Can't be elitist. [Laughter].[78]

Pushed, John Turnpenny admits at first that he himself can't see much difference between different types of urban open green space:

> I'd be interested to hear different people's views on the difference, what they think the difference is between commons and parks. Whether they think there is a difference […] Personally, I don't think there's very much of a difference. I'm sure this is the site of an enormous amount of debate. Personally, I think I certainly use them in the same way. They might be governed differently, or they might have different requirements, but in terms of me as a user, I feel similar about Mousehold as I do about Eaton Park, even though it looks very different.[79]

Later, after some consideration, John Turnpenny modifies his conclusion somewhat:

> what's the most important, is it that way that people interact with the space or is it the space itself? [...] It's varied. It's really varied. There are places here that aren't anywhere else, the heathland, the heather, the topography. Yeah, that's important.[80]

Asked whether he thought that the idea of "the commons" was generally contemplated by the users of the Heath, John Turnpenny replied:

> a lot of people would not really understand what commons were. If you said something like, 'The tragedy of the commons' people might think the House of Commons or common people or, I don't know, it's... the concept of common land I think is a little bit, it's not something that often gets talked about.[81]

Inextricably linked with ideas of *what* their urban open green space was, locals were as intrigued, if not more so, with *where* the edges of their park or commons were. Since the advent of ordnance survey mapping in 1745,[82] the need to establish and re-establish parochial and other local territorial boundaries through techniques of memory reinforcement like "beating the bounds" have receded.[83] "Beating the bounds" does persist in some areas, but now, rather than simply defending or memorialising an historic right to a place, it looks to the future; a practice now focussed on preserving places for the future, as some have argued.[84]

At Mousehold Heath, this boundary liminality is nicely captured in Trevor Nuthall's observation:

> you're in kind of two places [...] you're on the edge of what's being reclaimed and yet you're still in the wild area [...] it gives you the shape, I think. I like the idea of a shape of a space.[85]

Memorialisation is important here, but always looking to the future, Trevor Nuthall and Colin Howey in conversation reflected on this. Trevor Nuthall:

> we're walking here along a kind of processional way [...] there's a description of the boundary around here which is quite kind of fascinating [...] you think of people actually walking it to remind themselves and remind kind of younger people in the group what the boundaries were." Colin Howey: "Is this the beating of the bounds?" Trevor Nuthall: "Yeah." Colin Howey: "Parish boundaries.[86]

In contrast, others reported a kind of fluidity or at least an uncertainty to the edges of the Heath. Rosamund Chetleburgh admitted:

> I've no sense of direction. I used to get lost on here and one time popped out on the edge of the city and thought, 'Well, I had better walk home up Sprowston Road because here I am now about a mile and a half from home.'[87]

Inside, it seems, it is perceived as hard to find your way out and from the outside its edges are uncertain too. John Turnpenny, "unless you live not far from the Heath, it's very easy to not even realise that it's there".[88] Overlaid onto these shifting mental cartographies are other kinds of spatial mappings. Trevor Nuthall and Colin Howey again:

> the other element to this physical construction [...] is the construction of [a] sacred geography as well [...] going through increasing sanctity as you go through into that inner space [...] that sets up the expectation with those perimeters and boundaries. So, boundaries are really, really important in the spiritual realm as well as in the physical realm[89]

Moving Forward on Mousehold Heath

Mousehold Heath has a long history as a locus of contestation, as outlined in Chapters 3 and 7, but how do current users of the Heath feel about this and what is the Heath's future as a site in which to demonstrate grievance? Colin Howey's view is that:

> [...] it has the history of people coming together for [...] worship but also [...] to express their rights [...] And it is a commons [...] if you don't use and protect rights, they wither and go. So... I can't say that is, this certainly hasn't been a place for gathering for that, but I think we would lose that at our peril.[90]

Speaking specifically about his fears about climate emergency and how others might react in response to that, Colin Howey goes on:

> [...] it might well be expressing itself more in the ecological movements [now]. [...] I suspect [...] at the moment it's within the realms of the chattering classes but it might not be, because things can move very quickly. Once something actually becomes reality and people see it then perhaps we will see some kind of movement against it [...] I hope it's not lost and then we have to try and claw it back. It would be nice to stop it happening in the first place.[91]

Dwelling on the issue of rights to urban open green space, John Turnpenny argues that "rights and responsibilities are important. So, your right to... access, your right to enjoy the space, the responsibility to consider other people's enjoyment. I think quite simply, yeah".[92] Colin Howey is inclined agree, "I love the egalitarian aspect to this. This is a place of protest and dissent. I just love [that] it's on the margins".[93] Mousehold Heath and its most famous moment of protest during Kett's Rebellion, discussed at length in Chapter 7, is never far from the minds of some locals. Simon Floyd, Norwich theatre-maker, has staged a number of popular performances on the Heath centred on this event and dwells for a moment on the connections between people and place forged through situated performance, "giving people a sense of the history of a place connects people to that place. They love it when they hear their own place talked about – they feel a sense of pride and a sense of connection".[94]

Conversations about the future of urban open green space amongst those we interviewed inevitably, perhaps, turned to children and to young people in general. Health benefits were important to interviewees but in a more general way they were keen that their commons were on the one hand left in a fit state for future generations to enjoy, but also that children were sufficiently educated about those spaces so that they might become effective future custodians of them, "promoting opportunities for environmental leadership",[95] as Tanja Beer, Andrea Cook and Kate Kantor put it in an Australian context, as well as a need for "new approaches to environmental custodianship that celebrate children's agency".[96]

Rosamund Chetleburgh is clear that care for the Heath is about:

> it's keeping it going for future generations, especially the continuity here because all it's ever been is a heath, it's never been built on, it's not reclaimed, and that on the edge of a big city seems very important, to me. So, I'd like it to be here for future generations but I'd also like to try and be part, in some small way, of practically addressing it.[97]

John Turnpenny wonders for a moment about potential overuse of the Heath and whether smaller more local open green spaces would be a better solution:

> do we really want to encourage lots and lots of people to get in their cars and come halfway across Norwich, Norfolk to come and wander around on the Heath? Is it better to actually encourage people to use the places that are near them and have more but smaller spaces?[98]

Many understand, as John Turnpenny does, the value of the heathland itself, how unique it is and that this particular ecology is worth preserving, even nurturing and growing, for the future:

> I would like to see more of the heath part of it […] that rare habitat is really important, […] it's, a lot of the work, […] we're doing […] focused on keeping and expanding that area. I would like to see it just as green and certainly no smaller than it is now.[99]

In a less abstract way, Colin Howey saw the future wrapped up in the fate of the trees of Mousehold Heath:

> I love the idea that we can start that now for future generations to enjoy. […] Will [Mousehold Heath ranger] said to Gary, […], 'Gary, I would die for just

> one of these trees on Mousehold'. They're these amazing, gnarly old, pollarded oaks, hundreds of years old. And that's [...] And that's where they had that conversation [...] about creating the futures.[100]

Yet, Trevor Nuthall and Colin Howey were equivocal about how and whether younger generations would care, or know enough, to be adequate future custodians of the Heath:

> I mean I think they would see it as a great place of freedom, really." Colin: "Yeah, it would be nice if they did. [...] I think that probably youngsters just don't know it's here and don't engage with it as well. A lot of people in the city probably still aren't. [...] I suspect a lot of particularly youngsters maybe just don't know, don't connect. But that's something that could change. That's something that could be developed. [...] schools like Mile Cross Primary on the Mile Cross estate in Norwich have a Forest School area down there, so another natural step from that could be then to come up onto the Heath and do some stuff up here, for instance. So, there are ways that could be developed.[101]

But, in general, there was agreement that increased ability for as many and as diverse a group of people as possible to be able to visit was key, "I don't know how to make it more accessible, really. And it's time, isn't it"?[102] argued Nicola. As well as, according to John Turnpenny, "better public transport. Cheaper public transport".[103]

Little speaks to the future-facing aspects of urban commons as much as the opportunities they afford as learning spaces: opportunities for children and other young people, but also for adults. "Wastes and Strays" has come to understand urban green open space as a civic space and that these sorts of public space are an important component of "civic education".[104] The shift, for example, of children's outdoor play away from the street to the home and the resultant loss of learning opportunities has long been recognised. Alongside a better understanding of how landscapes might facilitate better educational and life outcomes, there has developed a more diverse understanding of open space as educational for children of all abilities.[105] Learning is, of course, essential for the young, but it is also a vital liberating force for all ages. Making intergenerational connections is seen as complex and important too. "I know people my age are quite hostile to youngsters", Paul Lynch admits, "and you think, 'Well, they're the ones who have got to be picking up our mantle, really'. If they're interested, they're going to make an effort".[106]

This sense of taking care of something, somewhere, to pass on to others is strong with the volunteers we spoke to, Trevor Nuthall emphasises that, "really, it's saying, 'Be careful of this area. Look after this area' [...] it's marking it as a special site".[107] There are also more straightforward child-orientated educational opportunities. Speaking about the Forest School she ran on the Heath, Nicola is sure that "those children feel connected with the place and then when they grow up, they'll want to protect it and keep it. [...] they value it".[108] Colin Howey muses more generally on the range of learning opportunities afforded by urban commons:

> there are so many different levels and layers you can engage people on a space like this [...] in terms of the ecology, the history [...] things like foraging and understanding that this is a living landscape that can help us live.[109]

But the challenges for stakeholders who would also be educators are profound. Colin Howey recalls:

> some of the audiences [...] I've brought up here, [...], over the years from former offenders, people in recovery from drugs and alcohol, people who have been homeless, [...] don't come up here necessarily thinking there's anything in it for them. They come up here thinking [...] 'So what?' so you've got to kind of prove it, prove yourself and prove it, share it [...] obviously there might be a few went away thinking, 'Well, that wasn't really for me', but I think everybody, particularly people who hadn't been here before [who] didn't know it was here, [...] felt that it's something a bit special.[110]

And what of the compulsion to build on our urban green open spaces? At times, in the descriptions of illicit den-building on Mousehold Heath, there are echoes of the "one night cottage myth" which cleaves especially to common land.[111] Tim Holt-Wilson explains:

> camp-building activity has rather been to the detriment of my geodiversity trail [...] Will does what he can to stop the camp building [...] every time I come here, it's been denned [...] he gets his volunteers, and they remove these bits of wood and they [...] scatter them around, hide them or whatever, [...] And then of course six months later you come back. 'Oh, they've done it again'. So, it's this endless dialogue between exposure of geology and construction of dens and encampments. [...] my policy is not really to complain or to insist on anything. It's just to accept that, you know, nature will find its way and one part of nature builds up a den and another part of nature knocks it down and scatters its materials. Dialectics of land use.[112]

On the other hand, architect and activist Guido Robazza insists that:

> involving citizens in proposing and evaluating spatial solutions for their city, but also producing the spaces as part of the process, can be an effective tool to raise awareness, inspire citizens and turn them into active agents of urban transformation.[113]

Even den-making on the Heath demonstrates that urban commons are perhaps latent arenas for creativity. Making structures, shelters, gathering places, even very simple ones, is a special case it seems of co-creative, participatory activity and is uniquely powerful as a method of connecting people to places.[114] Sometimes, liminal quality of commons can give rise to bathetic juxtapositions, "there was a homeless person camping there". Nicola recalls:

> but yeah, the kids, there's little dens that they make in the, I kind of cut a bit of the holly back so it was like more of a den. And up here as well there's another one in there. And it's just a really lovely thing to do.[115]

In terms of more systematic housing development, though, John Turnpenny, for example, is strongly opposed:

> I can imagine that there might be pressure to develop, new houses. I think should be resisted. It's very limited space, would be very easy to encroach on it bit by bit and then all of a sudden, you've lost maybe a third of it.[116]

Figure 9.2 A Manifesto for Mousehold Heath. A manifesto for Mousehold Heath devised with young adults at Sprowston Youth Engagement Project. The image is a still taken from the film created by youth project members Adam Davison, Aleksandr Derkac, Cora Kerrison and Nikola Kolus, in collaboration with Siobhan O'Neill and filmmaker, Annis Joslin. In-Common Sites: In Action. March 2022. Image by Siobhan O'Neill and Annis Joslin.

Urban Common Futures and Learning Through the Commons: Town Moor, Newcastle

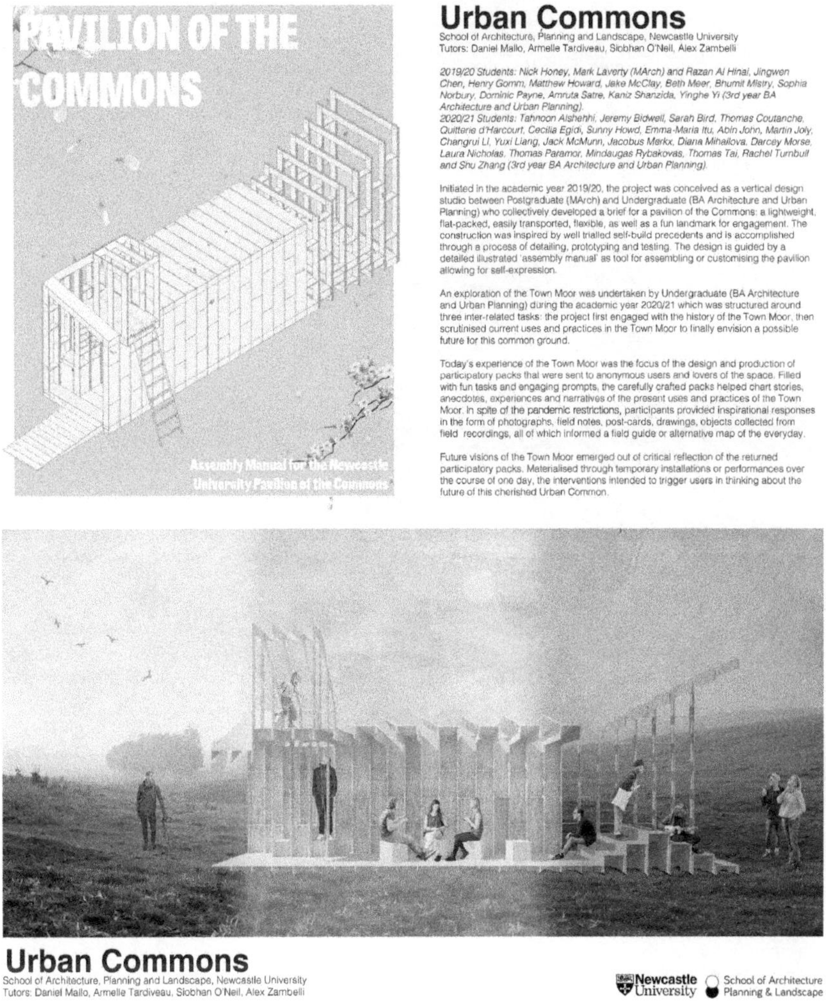

Figure 9.3 Pavilions of the Commons "Urban Commons" project by students of Newcastle University School of Architecture, Planning and Landscape, with permission Daniel Mallo and Armelle Tardiveau.

Figure 9.4 Live projects: top intervention in the public realm "Endangered Species" and bottom project "participatory pack" for cultural probe. Newcastle University School of Architecture, Planning and Landscape, with Daniel Mallo, Armelle Tardiveau, Siobhan O'Neill.

Urban Common Futures and Learning Through the Commons: Valley Gardens, Brighton

Figure 9.5 Pavilions of the Commons project by students of the University of Brighton School of Architecture, Technology and Engineering, with permission Graham Perring.

Figure 9.6 Pavilions of the Commons project by students of the University of Brighton School of Architecture, Technology and Engineering, with permission Graham Perring.

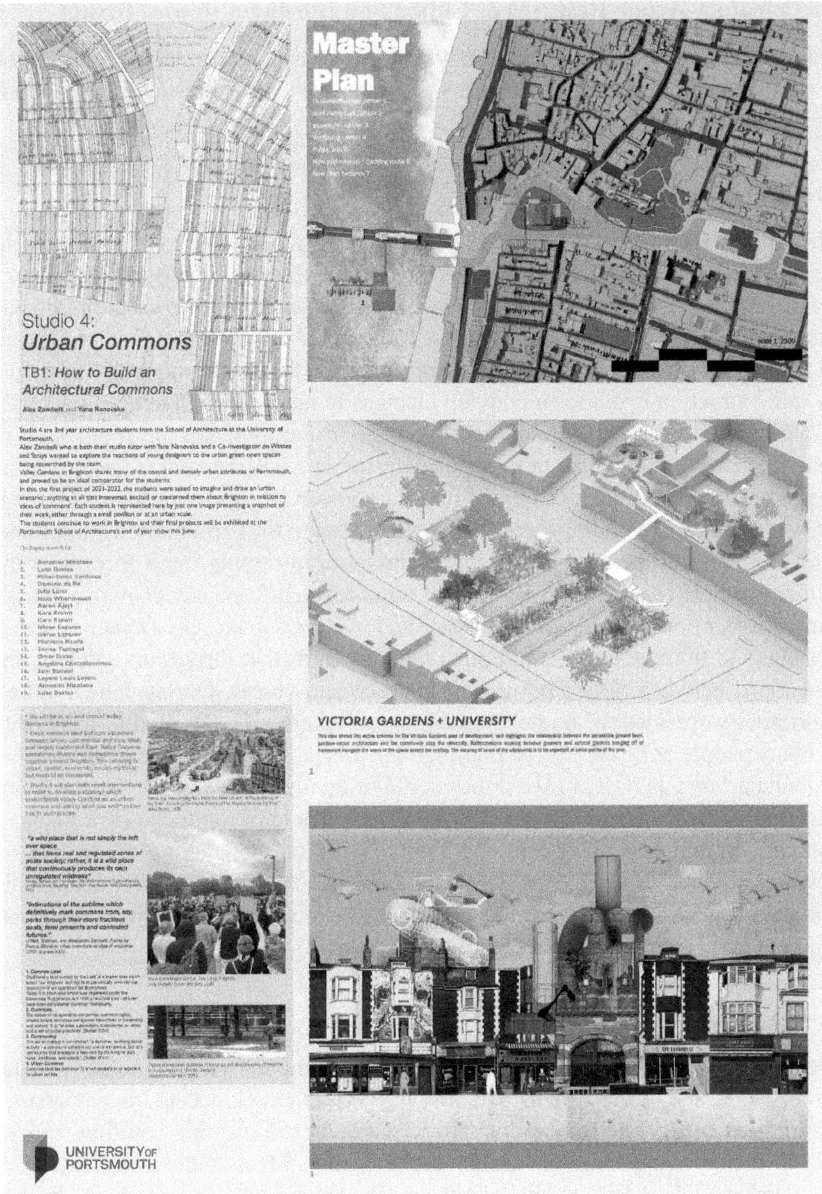

Figure 9.7 Pavilions of the Commons project by students of the University of Portsmouth "Urban Commons" Studio, Alessandro Zambelli and Yana Nanovska, 2021–2022.

Participatory Action Research at Clifton and Durdham Downs, Bristol

We set out here how Wastes and Strays project team member John Clarke developed a series of generally applicable writing co-creation exercises on Clifton and Durdham Downs.

Writing the Commons

Clifton and Durdham Downs have been shaped by our voices and by our writing. There is no part of them that has not been subject to some form of human desire, either to take something (minerals, quarry stone, timber, pleasure) or to add something (cricket fields, observatories, walks, viewing points, etc.). The Downs, as they are known in short, have in time been transformed from a place of industry and extraction to a site embodying changes in landscape aesthetics, spectacles of pleasure and punishment and now biodiversity farming and the stimulation of health and well-being. One thing that seems not to have changed is the Bristolian desire to slide down the rocks by Clifton Suspension Bridge, each successive generation of backsides adding their anonymous sheen to the carboniferous limestone, a feature that will perhaps prove one of the more intriguing fossil records of the human species. This complex layering of transient significance with more enduring expressions of power, parliamentary acts and bye-laws makes the Downs an exciting place to think about through poetry, an art form that thrives on quick-witted transitions between different kinds of material and voices. During John Clarke's short poetry "residency" on this popular urban common, he wanted to work with as many people as he could reach through the trusted local networks of our on-the-ground partners, in order to see if they could bring the complexity of the commons past and present into co-existence within the capacious commons of a poem, a comparison that only stands so long as they work on maintaining access and renewing ancient rights to the textual and textural riches of both. In order to maintain this double access, Clarke's residency was built around a series of workshops, set out here as an adaptable set of creative resources.

Postcards for the Downs

The first of these workshops explored the relationship between the visual and auditory landscape of the Downs. Clarke says he was struck during his preparatory exploration of the site by the dramatic acoustic transitions that could be experienced when moving from road traffic on the "Ladies Mile" to the alpine sounds of shattered rock dislodged by goats in the Gully; from sound systems in the bushes of the Dumps to the sounds of bees around a clump of green alkanet in a hollow where a rock had once been split by fire and extracted in manageable pieces; from the sounds of peregrines from the lookout over the Avon Gorge to the sounds of a drone and the tinkle of climber's carabiners on the cliff faces of the gorge. Here was a soundscape of acoustic complexity, unique to the multifarious and evolving use of the common, a present complexity that could be matched up with a complex history of visual representation in paintings, prints and postcards.

What participants were lacking in this equation were the sounds of the past. Clarke recalls being intrigued as to whether or not participants in a workshop might be able to reimagine and evoke in language the lost acoustic environment of the Downs. He attempted to do this through encouraging them to explore the transhistorical continuities of geological and natural sounds and changes in human-generated sounds that marked dramatic technological changes in our culture, for example, from horse's hooves to a Harley Davison, to the whine of a drone and a mobile phone ringtone. He was very fortunate to be connected by Mandy Lievers of the Avon Gorge and Downs Wildlife Project with the ex-curator of Bristol Art Gallery Francis Greenacre. Francis put together a handout of images of the Downs and guided the party from site to site around the commons as they attempted to locate the precise positions from which historic images had been made. At each site, the group stood in silence after their discussion, in order to allow themselves to be haunted by the body of the artist who had occupied the spatial point of view they now occupied. Doing this was a kind of spatial, performative archaeology that generated uncanny feelings that extended the experience beyond a simple comparison between then and now.

At significant places on the guided walk, the group stopped to do some writing and Clarke gave out prompts to get people writing (everyone had copies of the images to which Francis referred in his short histories of the site and the image). What follows is an adaptation of the workshop that could be used for any commons after a short period of research into its documentation in a library or archive:

1 Choose from a selection of images drawn from archives and local history books. Describe the soundscape implied by the picture, for example, cart wheels, voices and wind in the sails of a windmill. Focus on sounds in the foreground, middle distance and at the auditory horizon.
2 If possible, visit the location and make a note of the contemporary soundscape. In particular, think of how the man-made acoustic environment has changed.
3 Now write the sounds on a blank postcard or postcard-sized piece of paper, where the image would usually be. Think about the spatial arrangement of your words and how these might correspond to the image you're working with.
4 On the other side, address your postcard to someone or something in the archival image or imagine someone or something in the picture addressing the postcard to you in the future. Be inventive in your three-line postcard address. Think of how an address might direct you to think about unusual places and times on the commons. Be as specific as you can.
5 Think about the weather, atmosphere and drama of the archival image. If your image is of an empty landscape, think about what might have just happened or is about to happen. You might include information from the archive to help you with this.

6 With this research in mind, write a message to someone in the archival image or from the point of view of someone in the image. Think about the language used when writing postcards (or when texting); think how your tone changes according to your relationship to the person you're writing to. Whoever is writing the message must try to share a feeling or observation they don't feel able to say directly, for example, "I miss you"; "I feel ashamed"; "I can't stand this place".

What is particularly effective about using listening as a point of focus rather than vision is that it immediately defamiliarises the landscape: we often know what a place looks like, but it may become quite another place when we listen carefully to the sounds that shape its acoustic environment.

A Fair Field of Flowers

Clarke worked with two school groups from Westbury Park School on a glorious June day on the Downs: a group in the morning and a group in the afternoon. It was part of a whole day workshop during which, while the school groups were not working with him, he recalls, they were working with Mandy Lievers of the Avon Gorge and Downs Wildlife Project on finding out about the ecology of the Downs – from its wildflower to its Peregrines. On his part, Clarke wanted to explore how getting the children to use their words to touch and reveal the hidden or overlooked flowers in the wildflower meadow might open up a personal and playful imaginative possession of flowers and plants that caught their attention. And it worked. As one boy said to him during the workshop, pencil in hand, clipboard on knee: "I only like nature when I'm doing this". What they did was go through a list of activities that led up to a final performance of extraordinary invented flower names, a sort of wild florilegium of the imagination, the text for which have been included after this list of exercise prompts.

Here are Clarke's instructions to teachers wishing to replicate the workshop:

1 Working with the whole group, make a list of the simplified parts of a plant, for example, stem, flower, leaves, roots and, if you're feelings adventurous, the parts of the flower.
2 Find a plant close by that you're permitted to pick. Ask for a volunteer and work with them what is sometimes known as the "Furniture Game" – it's a simple and very effective technique for showing children they can make astonishing metaphors and connections between things in a moment and without thinking about it or rather thinking differently. You play the game by asking simple questions, for example, "If the stem were a kind of weather what would it be" or "If the flower were a song what would it be" or "If the roots were a kitchen utensil" what would it be? Encourage them not to censor themselves and celebrate the strange and unique connections they make. It's silly, it's fun and it's profoundly revealing about the way each child's mind works.

3 Now send the children off to play their own "Furniture Game" with the flower that catches their eye and note down the metaphors and similes they come up with. Encourage them to look so closely at the flower that they'll never ever forget it. Get them to gather a simile for each part of the flower.
4 Once they've done this, ask them to close their eyes and listen. Now ask them to listen as if they were the flower. What can the flower hear? Ask them to think about the different ways it might have listened and the different kinds of "ears" it might have, for example, roots, hairs and petals. After a couple of minutes or as long as they can manage, get them to open their eyes and immediately write down what they've heard as if they were the flower, that is, from the flower's point of view; they can use the first person or "I" voice to do so. Let the flower use similes to describe how they hear sounds. You might get the children to think about the kinds of things that flowers know about in order to arrive at effective similes. Ask them also to think about scale: how does being small change your relationship to the sound of a dog, a football being kicked, etc.
5 Get the children to read out one of the things they've heard and create a collaborative poem either by writing them down or recording them using a phone.
6 Now get them to imagine the flower is either an old-fashioned telephone, a mobile phone, an ear trumpet or simply an ear. Who might be listening at the other end of the flower phone? What might you say to them and how might you say it? There are many choices and none of them right or wrong. Encourage the children to play around with the idea.
7 Bring all these notes and thoughts together in a simple form of guided writing that involves using the writing already made to complete the following sentences. They are going to talk and listen to their plant!

i. Describe a part of the plant
ii. Complete the sentence: "I look into your ….. and see….."
iii. Complete the sentence: "I listen to your ….. and hear….."
iv. Ask it a question
v. Give the flower's response
vi. From the flower's point of view, complete the following: "At night/dawn/noon, etc. I can hear …..".
vii. Invent a compound name for the flower using nouns from three similes generated in (2) and complete the following sentence: "The say your name is [give the usual name] but I call you [give the new name made of nouns from the simile game]".

You can improvise with these exercises depending on how much time you have. Here's an example of the collaborative poem each group produced that combined all the names the children had made up for their respective flowers. It's a joyous and crazily inventive list that undermine conventional flower taxonomy, and my hope is that each child will always remember their secret special name for their flower.

A Field of Flowers #1

Summer lion
Lank-fruit lark
Pudderhead flaming asparagus
Pluso-melon
Orange web
Wrist fire
Venus ingot
Puffer pillow-sun
Lullyingus
Summer ruby
Teddy-bear spring-rains blossom tree
Pom-pom
Low magnolia
Willy-earring
Snowy pop flake
Sunny messy knife
Early milk magnolia
Peacock lightning
Purple wormhole
Cobweb satin toffee
Foxtail fire-muscle
Saffron egoslavia
Earth chin-bag
Puffball mars
Milky way
Little ear wiggy piggy
PCP Z47
Blank shard

Westbury Park School, Bristol, The Downs, June 2021

A Field of Flower #2

Pineapple rocket
Soft death
Snowy moon pearl
Lilac heather
Sunny stag beetle
Calm butterfly's yoga
Blind popcorn
Floral Mango sunset
Lightning wedding dress
Sweet evening sky
Blue flam
Sunny buzzing bee
Stripy tiger floor lamp
Citrus sunset
Marshmallow cloud
A ghost snowball
Green grumpy coal-flower
Orange-juice shorts
Under-bun
Summer lamp
Midnight spaghetti onesie
Snowball
Pink puffy fish
Berry planet
Purple fire
Bob jnr jnr
Lonely ghost
Lemon cappuccino
Marwacker
Mould mack

Westbury Park School, Bristol, The Downs, June 2021

Notes

1 See Efthalia Thaleia Grigoriadou, 'The Urban Balcony as the New Public Space for Well-Being in Times of Social Distancing', *Cities & Health* (2020). Mark Scott, 'Covid-19, Place-Making and Health', *Planning Theory & Practice* 21 (2020), and Vikas Mehta, 'The New Proxemics: Covid-19, Social Distancing, and Sociable Space', *Journal of Urban Design* (2020).

2 Martin Joly, *'Intervention Responses' Newcastle University School of Architecture, Planning and Landscape, Student Project Intervention. Probing the Future* (Newcastle University School of Architecture, Planning and Landscape, 2021).

3 Aleksandr Derkac, *Participant Contribution, in-Common Sites in Action Workshop*, (8 February, 2022).

4 See 'C40 Cities', (C40 Cities Climate Leadership Group, 2022). and 'What Is Levelling up and Who Is It Helping?' (BBC, 2023).

5 David Harvey, 'The Right to the City', *New Left Review* II (2008), 23.

6 See Joel Golby, 'We Are All in the Gutter – but Some of Us Are Counting the Nitrous Oxide Canisters', *The Guardian* [online] (2020), and Kate Power, 'The Covid-19 Pandemic Has Increased the Care Burden of Women and Families', *Sustainability: Science, Practice and Policy* 16 (2020).

7 *Walking Conversation with Tim Holt-Wilson* (26 June 2022).

8 A concept which gained currency after Peter Linebaugh's popularisation of the term; Peter Linebaugh, *The Magna Carta Manifesto: The Struggle to Reclaim Liberties and Commons for All* (Berkeley: University of California Press, 2008).

9 Dr Siobhan O'Neill instigated, developed and led the 'In-Common Sites', 'In-Common Walking' and 'In-Common Sites in Action' activities and collaborated in the Newcastle University 'Urban Commons' student projects with Daniel Mallo and Armelle Tardiveau.

Dr Alessandro Zambelli instigated, developed and led the 'Pavilions of the Commons' activities as well as the 'Urban Commons' University of Portsmouth student projects.

Dr John Clarke instigated, developed and led the 'Writing the Commons' activities.

10 F. Baum, C. MacDougall, and D. Smith, 'Participatory Action Research', *J Epidemiol Community Health* 60 (2006).

11 Early events were held online and as local COVID-19 lockdown rules relaxed we were able to meet participants in person.

12 See, for example, Guido Robazza, 'Build Art, Build Resilience. Co-Creation of Public Art as a Tactic to Improve Community Resilience', *The Journal of public Space*, 5 (2020), or Christina Horvath and Juliet Carpenter, *Co-Creation in theory and practice: exploring creativity in the global north and south* (Bristol: Polity Press, 2020).

13 Extract adapted from their design studio brief, Daniel Mallo and Armelle Tardiveau, 2020.

14 Pedagogical demands meant that this had to be a quicker version of the project.

15 D. R. Denman, 'The Future Use of Common Land', *Journal of the Royal Society of Arts*, 112 (1964).

16 Ibid., p. 232.

17 Katrina Myrvang Brown, 'New Challenges for Old Commons: The Role of Historical Common Land in Contemporary Rural Spaces', *Scottish Geographical Journal* 122 (2006): 109.

18 Johan Colding and others, 'Urban Green Commons: Insights on Urban Common Property Systems', *Global Environmental Change* 23 (2013): 1049.

19 *Walking Conversation with John Turnpenny* (25 June 2022).

20 *Walking Conversation with Owain Hall* (13 September 2022).

21 *Walking Conversation with Paul Lynch* (26 June 2022).

22 *Walking Conversation with John Turnpenny*.

23 Jack Anderson, 'Village Greens, Commons Land and the Emergence of Sports Law in the UK', *The International Sports Law Journal* 14 (2014), introduction.
24 Ibid.
25 Katherine King and Andrew Church, 'Beyond Transgression: Mountain Biking, Young People and Managing Green Spaces', *Annals of Leisure Research* 23 (2020): 203.
26 Ana Morcillo Pallares, 'Hold, Denote or Reserve: Spatial Strategies in Detroit's Available Space', *104th ACSA Annual Meeting Proceedings*, Shaping New Knowledges (2016).
27 *Walking Conversation with Paul Lynch.*
28 *Walking Conversation with Nicola* (27 June 2022).
29 Ibid.
30 This would have been because of lockdowns caused by the COVID-19 pandemic.
31 *Walking Conversation with Nicola.*
32 *Walking Conversation with Rosamund Chetleburgh* (27 June 2022).
33 Ibid.
34 Andrey Egorov I. and others, 'Urban Green Spaces: A Review of Evidence', ed. by European Environment and Health Process (World Health Organisation: Europe, 2016).
35 Ibid.
36 C40 Cities.
37 *Walking Conversation with John Turnpenny.*
38 Usually translated as "it is solved by walking", that is, by experiment. *Walking Conversation with Trevor Nuthall and Colin Howey* (29 June 2022).
39 Ibid.
40 Ibid.
41 *Walking Conversation with Rosamund Chetleburgh.*
42 *Walking Conversation with Tim Holt-Wilson.*
43 For an overview of the provenance of and difference between these terms, see Melissa Anne Beryl Vogt, 'Agricultural Wilding: Rewilding for Agricultural Landscapes through an Increase in Wild Productive Systems', *Journal of Environmental Management* 284 (2021): 6.
44 Michael Soulé and Reed Noss, 'Rewilding and Biodiversity: Complementary Goals for Continental Conservation', *Wild Earth* 8 (1998).
45 See https://www.rewildingbritain.org.uk, https://rewildingeurope.com, https://rewilding.org and https://www.wwf.org.au/what-we-do/rewilding-australia
46 *Walking Conversation with Robert Stubbs* (17 August 2021).
47 *Walking Conversation with Trevor Nuthall and Colin Howey.*
48 Ibid.
49 *Walking Conversation with Paul Lynch.*
50 *Walking Conversation with John Turnpenny.*
51 *Walking Conversation with Trevor Nuthall and Colin Howey.*
52 Ibid.
53 *Walking Conversation with Paul Lynch.*
54 *Walking Conversation with Trevor Nuthall and Colin Howey.*
55 *Walking Conversation with Nicola.*
56 Ibid.
57 Ibid.
58 Ibid.
59 *Walking Conversation with John Turnpenny.*
60 *Walking Conversation with Rosamund Chetleburgh.*
61 A. Barker et al., *Understanding Green Infrastructure: A Signposting Guide* (Manchester: University of Manchester, 2019).
62 Cliff Moughtin and Peter Shirley, *Urban Design: Green Dimensions*, 2nd edn (Oxford: Architectural Press, 2005), p. 73.

63 Walking Conversation with Trevor Nuthall and Colin Howey.
64 Walking Conversation with Robert Stubbs.
65 Christopher Short and Michael Winter, 'The Problem of Common Land: Towards Stakeholder Governance', *Journal of Environmental Planning and Management* 42 (1999), 614.
66 Walking Conversation with John Turnpenny.
67 Ibid.
68 Walking Conversation with Trevor Nuthall and Colin Howey.
69 Walking Conversation with John Turnpenny.
70 Ibid.
71 Walking Conversation with Rosamund Chetleburgh.
72 Lindsay Campbell and Anne Wiesen, *Restorative Commons: Creating Health and Well-Being through Urban Landscapes* (Philadelphia, PA: US Department of Agriculture, Forest Service, Northern Research Station, 2009), p. 12.
73 Walking Conversation with Trevor Nuthall and Colin Howey.
74 Ibid.
75 Walking Conversation with John Turnpenny.
76 Garrett Hardin, 'The Tragedy of the Commons', *Science* 162 (1968).
77 Walking Conversation with John Turnpenny.
78 Ibid.
79 Ibid.
80 Ibid.
81 Ibid.
82 Rachel Hewitt, *Map of a Nation: A Biography of the Ordnance Survey* (London: Granta, 2010).
83 Amelia Soth, '"Beating the Bounds": How Did People Find out Where Their Local Boundaries Were before There Were Reliable Maps?', in *Cabinet of Curiosities* (JSTOR Daily Online, 2020).
84 Paul Oliver, '"Beating the Bounds": Switching Boundaries over Five Millennia', Traditional Dwellings and Settlements Review, 15 (2004), and Soth, "Beating the Bounds".
85 Walking Conversation with Trevor Nuthall and Colin Howey.
86 Ibid.
87 Walking Conversation with Rosamund Chetleburgh.
88 Walking Conversation with John Turnpenny.
89 Walking Conversation with Trevor Nuthall and Colin Howey.
90 Ibid.
91 Ibid.
92 Walking Conversation with John Turnpenny.
93 Walking Conversation with Trevor Nuthall and Colin Howey.
94 Walking Conversation with Simon Floyd (22 November 2021).
95 Tanja Beer, Andrea Cook, and Kate Kantor, 'Running Wild: Engaging and Empowering Future Custodians of Place through Creative Nature-Based Play', *Journal of Public Pedagogies* 3 (2018), 5.
96 Ibid., p. 6.
97 Walking Conversation with Rosamund Chetleburgh.
98 Walking Conversation with John Turnpenny.
99 Ibid.
100 Walking Conversation with Trevor Nuthall and Colin Howey.
101 Ibid.
102 Walking Conversation with Nicola.
103 Walking Conversation with John Turnpenny.
104 Ash Amin, 'Collective Culture and Urban Public Space', *City* 12 (2008): 10.

105 See Gill Valentine and John McKendrick, 'Children's Outdoor Play: Exploring Parental Concerns About Children's Safety and the Changing Nature of Childhood', *Geoforum* 28 (1997); Brenna Hassinger-Das and others, 'Learning Landscapes: Playing the Way to Learning and Engagement in Public Spaces', *Education Sciences* 8 (2018); Victoria Derr, 'Urban Green Spaces as Participatory Learning Laboratories', *Proceedings of the Institution of Civil Engineers - Urban Design and Planning* 171 (2018); and Nadia von Benzon, 'Unruly Children in Unbounded Spaces: School-Based Nature Experiences for Urban Learning Disabled Young People in Greater Manchester, UK', *Journal of Rural Studies* 51 (2017).
106 *Walking Conversation with Paul Lynch.*
107 *Walking Conversation with Trevor Nuthall and Colin Howey.*
108 *Walking Conversation with Nicola.*
109 *Walking Conversation with Trevor Nuthall and Colin Howey.*
110 Ibid.
111 Danae Tankard debunks the myth promulgated, she argues, principally in the work of Sayce and Gill; Danae Tankard, 'The Regulation of Cottage Building in Seventeenth-Century Sussex', *The Agricultural History Review* 59 (2011); R U Sayce, 'The One-Night House and Its Distribution', *Folklore,* 53 (1942); Gill W. W., 'The One-Night House', *Folklore* 55 (1944).
112 *Walking Conversation with Trevor Nuthall and Colin Howey.*
113 Robazza, 'Build Art, Build Resilience', p. 299.
114 Horvath and Carpenter, *Co-Creation in theory and practice*; Elizabeth B. N. Sanders and Pieter Jan Stappers, 'Co-Creation and the New Landscapes of Design', *CoDesign* 4 (2008).
115 *Walking Conversation with Nicola.*
116 See Figure 3.2.

10 Whispers of Better Things – Re-Imagining the Future of the "Urban" Common[1]

Urban Commons, Common Themes

The research presented in this book has focused attention on the importance of our surviving urban commons. Each urban common is unique. And if we are to understand their importance to local communities, we need to engage with their past and consider how their individual histories and development have shaped their community's sense of identity, belonging and "ownership" of the space. The case studies presented in Chapters 2–5 illustrate these points very clearly. At the same time, there are several broad themes that are common to all urban commons and that our case studies also manifest. The first is a disconnect between (i) the "formal" (or legal) classification of urban commons, especially that represented in property rights, and (ii) community attitudes to "ownership", identity and belonging. A second theme concerns public attitudes to the relevance (or otherwise) of long usage of the space to creating community "rights" for the public to use it. And it is often assumed that urban commons have survived unchanged for centuries, whereas change has, in fact, been constant, as demonstrated by the case studies presented in this work – both in the physical attributes of the common itself and also in the shifting and changing community uses to which they have been put from time to time. We will conclude in this chapter by considering some of these emerging problems and also by considering the position and role of urban commons as a distinct category of green "space" in developing innovative public policy for the future.

The disconnect between formal and community conceptions of urban commons becomes somewhat clearer if we think in terms of Henri Lefebvre's theoretical analysis.[2] This posits a disconnect (or divergence) between the urban common as "conceived" space and urban commons as "lived" (or what Lefebvre calls "representational") space.[3] By conceived space we mean here the conceptualised space of scientists or planners reflected in the pre-existing legal order for the regulation and use of common space. This will differ from "representational" space – space as directly lived through its associations with images and symbols and hence the space of inhabitants and users. The second broad theme – the relevance or otherwise of long public use of open space – is represented in the legal concept of prescription as a mode of creating property rights in individuals and/or the community. This was, for example, an active issue in the Victorian litigation over Mousehold Heath

DOI: 10.4324/9781003204558-13

considered in Chapter 3, and is also relevant to the ability of communities through long communal practice to "create" new commons for future community use – or to at least have their "rights" recognised in legally sanctioned property rights. We will explore these themes below but will first consider the developing public policy agenda for green space – and the relevance and role of urban commons within that discussion.

Public Policy for Green "Space" – What Role for Urban Commons?

The COVID-19 pandemic highlighted the need to make sufficient "green" space, including commons, available in urban areas. The public health measures introduced during the pandemic, especially restrictions on travel, work and social interaction, have demonstrated the need for suitable urban green space that can be safely accessed for open-air recreation. Increasing our access to "urban green spaces" could not only deliver recreational benefits, but also offer significant improvements in physical and mental health. Its recreational benefits are, furthermore, enhanced by people being directly exposed to nature.[4] "Green exercise" studies have shown that while people already benefit from spending time in green space, contact with nature significantly improves psychological health.[5] And it has been shown that exposure to natural *landscapes* while undertaking green exercise enhances its psychological and physical health benefits.[6] There are also huge potential resourcing benefits – it has been estimated that increasing the use of urban green space could generate a reduction in health expenditure of £2.1 billion per annum.[7] Its ability to "reconnect" people with nature is just as important as the recreational value of urban green space; enhancing their biodiversity would also optimise the cultural ecosystem services that our urban commons can generate.

Nevertheless, urban green space is under threat. Cuts to local authority budgets has meant that selling off publicly owned land is an easy option for public bodies struggling to maintain their front line community services. In 2018, for example, the charity *Locality* published research showing that on average, 4,131 public buildings and parks are being sold off each year by cash-strapped councils.[8] And planning policy continues to reflect a relentless drive for more housing development. Publicly owned green space in urban areas often offers a tempting quick fix for local authorities facing demanding targets for delivering new housing.

There are, however, encouraging signs that the need to increase not just the amount of green space in our urban areas, but also its accessibility to local populations, is at last being recognised in public policy. The government's Environmental Improvement Plan 2023 includes, for the first time, a commitment to provide access for the public to green space or water within a 15-minute walk from their home.[9] This is to be supported by an *Access for All Fund* of £14.5 million and by funding through the *Levelling Up Parks Fund* to create or refurbish more than 100 green spaces. These are lofty statements of ambition, but delivering them will be challenging.

The question today is, of course, *how* do we save and/or improve our existing urban green spaces and how do we "increase" our urban open space? And how do we assess (or measure) its utility to local communities? As we have seen in the case studies presented in this work, our urban commons provide a unique and valuable form of green space which merits substantial further investment. But *how* do we create *new* urban commons? And how do we assess the relative value of urban commons against other categories of urban green space when considering the merits of public investment in their future? The legal terrain is complex, with public parks, allotments, village greens and urban commons all manifesting different origins, legal classifications, property regimes and management structures. The rules for their creation, management and governance also vary in each case. Recognition of the need for greater investment in our green spaces is a welcome development, but implementing an expansive strategy for delivering new green space will face significant challenges. What are the problems that we face in recalibrating our policy and attitudes in order to protect and promote urban green space, what can be done and how?

Classifying Urban Commons – A New Approach?

If we are to improve and expand the green space available to our urban communities, then we will need to carefully consider, and possibly rethink, the classification and terminology used to describe and define urban green space. We currently have a number of categories of different types of urban green space. These include, for example, urban "commons", town and village greens (TVGs), allotments, and public parks, each with their own legal definitions and rules and in each case using differing terminology. This is arguably unnecessary, outdated and confusing. Why?

Take first the "urban common". What we typically refer to as "urban commons" are multifaceted spaces which all perform similar social and ecosystem functions. But their legal classification is often misunderstood, as is their protected status (or in some cases, their lack of legal protection). They share key characteristics:

- All are spaces subject to multiple uses: for example, recreation, sports, meeting spaces and allotments.
- Different "commons" have different origins. Many originated as "waste" of a manor – land that was unproductive and left for residents to use for grazing livestock or growing essential crops or vegetables. These may have originally been on the periphery of urban areas or wholly rural in nature, but have become "urban" commons where settlements have grown up around them. Mousehold Heath in Norwich is a case in point – originally waste of the manor of Pockthorpe and outside the city, it was grazed by livestock and administered by the Pockthorpe manorial court (overseen by the Dean and Chapter of Norwich Cathedral as lords of the manor). It is now fully within the urban boundaries of Norwich and provides a vital "green lung" for the city.
- What are typically referred to as urban "commons" in fact have differing legal classifications. Some are commons and registered under the Commons Registration Act 1965; some may, in fact, be registered as TVGs; some may be held as public parks; and yet others may have a bespoke status as open green space

for community use, a status guaranteed by statute (as in the case of Town Moor in Newcastle, which has been protected by statute since 1774).[10]

- Different "commons" have differing levels and types of legal protection from development, sale and disturbance. The strongest protection is found where an area of public green space is protected by a private Act of Parliament, as is the case, for example, with Town Moor in Newcastle, Mousehold Heath in Norwich or Clifton and Durdham Downs in Bristol.[11] Green space that is simply designated as a public park enjoys less protection from development.[12]

The legal classification of urban commons is therefore fragmented. It is also very poor at capturing the multiple uses of green space and the different ecosystem services that it supplies in a modern context. It tends to focus on a specific (and in each case different) ecosystem service provided for each type of urban "common". Moreover, the legal definition of some types of urban "common" is "bounded", that is, their definition reflects a limited provision of ecosystem services to specific users or groups of common user. So, in the case of an area of green space that is classified in law as a *common* and registered as such under the Commons Registration Act 1965 or Commons Act 2006, the focus of land use was originally on resource use (typically agriculture), but today, the principal ecosystem service that it provides in an urban context will be recreational public access. Access to the common for recreational purposes is not bounded – open access is guaranteed by the Countryside, etc. Act 2000[13] or under the terms of specific statutes.[14] If, on the other hand, green space is registered as a *Town and Village Green*, then this provides only for recreational use (access for "lawful sports and pastimes"). And a TVG gives only bounded access, that is, to residents of a "neighborhood within a locality" where the TVG is situated. *Public Parks* give access for solely recreational use, but this is usually not bounded – anyone can use the space. *Private Use Rights* can subsist as easements or profits, but these are strictly bounded (they will only give access to the legal "owner" of the property rights). Finally, if land is designated in a local development plan as *Local Green Space* it will give unbounded access. This is a planning designation for land demonstrating "beauty, historical significance, recreational value, tranquillity or the richness of its wildlife".[15]

In a contemporary context, these typologies increasingly lack relevance and purpose. The one attribute that all of these different categories of urban green space share is that they all provide vital ecosystem services from which we benefit. These change over time, as does their value to the community, their role in defining "place" and identity and their environmental significance. They may, for example, provide *resources* for industry and/or agriculture (these are often reflected in resource use rights such as estovers, grazing rights); *recreational access* and places for sport and recreational pastimes (an important cultural ecosystem service); spaces for social and political gatherings, and for protest; and many contain or represent sites of *cultural heritage.* The oral history fieldwork presented in the case studies in Chapters 2–5, is itself illustrative of changes in the use of urban commons, their importance and their relevance to communities.

Rather than focusing on the outdated and confusing legal classification of green space that we have inherited (outlined above), we need to revisit our understanding of green space – its importance and functions – to reflect contemporary priorities and needs. In other words, we need a new *functional* definition of urban commons and other "green" space. This should focus on the ecosystem services that contemporary urban commons provide for the community rather than on their origins and legal definition. It should also focus attention on the much-needed ecosystem services that designating *new urban commons* could provide.

Property, "Ownership" and the Urban Common

A key characteristic of the urban common – one that demarcates if from other forms of urban green space, such as a village green – is the strong sense of community identification and "ownership" that it engenders in members of the community. In some cases, this is attributable to the existence of formal common property rights vested in members of the community. These might be estovers, for example, the right to take berries, fallen fruit or timber. Or there may be a right to graze livestock on the land. But the creation of a strong bond (or sense) of community ownership can also be associated with long usage of the land for open access, recreation and other communal activities. Of the case studies in this work, for example, only one – Clifton and Durdham Downs – today manifests formal property rights vested in members of the community, in that case registered common rights to graze sheep on the Downs. But these rights are only infrequently, and symbolically, exercised. Yet the oral history research outlined in Chapters 2–5 shows that in all four case studies a strong sense of community "ownership" has developed, and this is closely related to a sense of the urban common's role in shaping the identity of the community. As noted above, this illustrates the divergence of the conceived (or formal) sense of the common as shared space, and the "lived" reality of the common as communal space. Another feature of the urban common that makes it unique and separates it from other forms of "green space" is the importance of "commoning" – shared practice and use of the space. These aspects of common practice were explored in Chapter 9. And, as we noted in the Introduction to this book,[16] the notion of "space" as opposed to place is that *space is a practiced place*.[17] The notion of an urban common as a green space is therefore dynamic and reflects the uses made of it, for example, as recreational space or for promoting community gatherings for protest or sports.

How then are we to promote the development of new urban commons for the future and how are we to develop new imaginaries for our urban commons? A new approach might focus on identifying and promoting the ecosystem services provided by urban commons (whatever their precise legal classification), and this could be the centrepiece for a *Community Charter for Urban Green Space*. This would fulfil two key objectives. In the first place, we need to ensure that existing urban commons, village greens and public parks are protected from development and maintained as open spaces for public recreational use. And second, we need to create new "green space" in cases where none is currently available in urban and peri-urban areas for open access by the public. Increasing the legal protection afforded to existing village greens and commons – and to land that is informally used for community recreational purposes – would require a reversal of the policy reflected in recent legislation, which has favoured development over public rights of recreational access. There is constant pressure on "protected" green space, with regular applications to deregister village greens and areas of common land for development.

One means by which communities can assert legal control over the protection of green space is by seeking to get land that has been used for sports and/or recreational pastimes registered as a town or village green ('TVG') under the Commons Act 2006. Land can be registered, where, for not less than 20 years, a "significant number of the inhabitants of any locality, or any neighbourhood within a locality" have indulged in lawful sports and pastimes on the land.[18] Landowners can also voluntarily dedicate land as a TVG under provisions introduced by the Commons Act 2006. A landowner could also dedicate a "new common" (rather than a TVG) by creating common rights vested in individual members of the community; this will secure registration of the land as "common" land with open public access under the Countryside and Rights of Way Act 2000. And individual common rights can be created by prescription, that is, use for 20/40 years to create a *profit a prendre.*[19] This could, in principle, also lead to the creation of a new "common" with public access rights under the Countryside etc. Act 2000.

Communities have had some success with TVG registration over the last 20 years or so, aided by a generous interpretation of the law by the courts. It has been held that, for example, land need not conform to a specific physical description in order to be a TVG: it does not have to represent the traditional village "green" and may include part of a beach or even land that has been "created" by the deposit of soil.[20] And once registered, a TVG can be used for all "lawful sports and pastimes".[21] In the *Sunningwell* case,[22] the House of Lords somewhat generously held that "lawful sports and pastimes" can include a wide range of recreational activities, including dog walking, playing with children, blackberry picking and other forms of informal recreation. This sparked an uplift in applications for the registration of land as TVGs, which would then be available for a wide range of community uses.

Community-led action to secure green space suffered a blow when the Growth and Infrastructure Act 2013 blocked applications to register a TVG when an application for planning permission to develop the land had been made or where land is allocated for development in a local development plan.[23] The effect of this is to favour planning policy to maximise house building at the expense of protecting green space for existing communities. And individual common rights can no longer be acquired by prescription.[24] This will preclude the creation of a registrable common through long use of resource access rights, with associated public access rights under the "right to roam".[25]

The judiciary has also played a part in restricting community rights to acquire access rights over land by limiting their ability to secure the registration of TVGs.[26] The Supreme Court gave two rulings in 2019 that greatly inhibit community rights to secure TVG registration. The first was an appeal by Lancashire County Council (as local education authority) against the registration of 13 hectares of land in the "Moorside Fields" case. The second was an appeal by the NHS against the registration of 2.9 hectares in the "Leach Grove Wood" case. They ruled by a 3:2 majority that where land is held by a public body for statutory purposes which are inconsistent with the recreational rights conferred by TVG registration, then it cannot be registered. This was based on the principle of statutory incompatibility – would the recognition of community recreational rights over a TVG make it impossible for a public body to carry out the other statutory purposes for which it held the land? In an earlier case, the Supreme Court had applied the statutory incompatibility principle to rule against the registration as a TVG of a specific area of land held by the Newhaven Port Authority for the purposes of its port operation.[27]

The decision in the Moorside Fields case greatly extends this principle to deny registration as a TVG to land that is held by public bodies which have *general* powers granted by statute to hold *any* land for educational or health purposes.[28] Incompatibility is to be decided by reference to the statutory purposes for which the land is held by a public body and not by looking at the actual *use* of the land at any given time.[29] The land here could not, therefore, be registered as a TVG, even though its sole intended future purpose was to provide outside activities and sports for the school – uses which were not necessarily incompatible with the use of the land by local inhabitants for lawful sports and pastimes.[30] This decision will greatly restrict the ability of local communities to register as TVGs any land that is owned by public bodies – and to secure its long-term protection as community green space.

Towards a New Community Charter for Urban Commons

The best response to the pressing need for greater access to green space – a need highlighted by the pandemic – would be the introduction of a *Community Charter for Urban Green Space*, a new charter focused on identifying and capturing the ecosystem services provided by urban commons and other forms of green space. Requiring local authorities to carry out an *ecosystem services appraisal* of areas of green space could be a powerful tool in protecting urban commons for the future: identifying user groups, the numbers using urban green spaces for different purposes and then scoping its potential for the development of new ecosystem services for our communities. This could go hand in hand with a review of the legislative provisions currently governing the various categories of urban common and other types of green space.

The statutory power to voluntarily dedicate land as a TVG might be useful where public bodies – for example, the NHS – wish to create new green space to secure the community health benefits it can deliver. Similarly, a landowner could dedicate a "new common" by creating common rights vested in individual members of the community – this will secure registration of the land as "common" land with open public access.[31] This might involve, for example, vesting rights of estover in members of the local community – the right to take the natural produce of the land, such as a right to pick fruit from trees, or to collect wood or berries in autumn.[32]

The benefit of this approach would be to invest the community with property rights in the land and engender a sense of community "ownership" and stewardship. The key problem in both cases, of course, will be persuading landowners to dedicate land for these purposes. And community rights to secure registration based on long customary usage should be reviewed and reinforced. Consideration should be given to repealing the 2013 Act provisions barring registration of TVGs where planning consent has been applied for, and to introducing legislation to reverse the negative impact of the Moorside Fields case on a community's ability to seek registration of land held by public bodies as a TVG.

We also need to refocus the planning system to promote green space. First, we need to review the development planning process to prioritise the provision of community green space. The setting aside of land as village greens or new commons in local development documents would be an important first step.[33]

We also need to review the rules for designating "local green space" in local and neighbourhood plans. Local planning bodies have the power to designate "local green space" in local and neighbourhood plans, but this power is heavily circumscribed in current planning guidance. This stresses the need for it to be "demonstrably special" to a local community, for example, because of its beauty, historical significance, recreational value, tranquillity or the richness of its wildlife.[34] The local green space designation should also not be used for "extensive tracts" of land, and is qualified, in that it should "complement" investment in sufficient homes, jobs and other essential services.[35] This is a poor substitute for the protection of recreational access rights through the registration of a TVG or new common. Nevertheless, carrying out an ecosystem services appraisal, as suggested here, would be a useful tool in establishing the "demonstrable" importance of land for the purposes of the Local Green Space planning designation. It should also not be forgotten that other existing planning tools, such as planning agreements and the community infrastructure levy, can be used to secure funds for the provision of new recreational space, including urban commons, in our urban communities. Existing planning powers need to be proactively used to this end by public bodies and supported by much stronger planning policy guidance to promote both the creation and protection of urban commons and (where appropriate) other forms of recreational green space.

If Octavia Hill's vision is to become a reality, we must find a way to refocus planning rules and planning guidance to promote green space as an integral and essential element of all new development and create new urban commons where there is currently a lack of available green space within easy reach of urban communities. As suggested here, a key change in this approach should be a new focus on the ecosystem services that urban commons provide. Introducing an *ecosystem services appraisal* as an integral element of a new Charter for Urban Green Space would be a major step forward. This could be used in new planning guidance for local green space, would be a material consideration in development control decisions on all new development and would help to refocus attention away from the technical legal classification of different types of urban green spaces – and towards a more functional approach highlighting and valuing their role as a key source of ecosystem services for our communities.

Notes

1 "*Whispers of better things come to us gently*" (Octavia Hill, 1883).
2 See further Chapter 1, p 3 above.
3 Henri Lefebvre, *The Production of Space* (tr. Donald Nicholson-Smith, Oxford: Blackwell, 1991), pp. 38ff.
4 Jules N. Pretty, Jo Barton, Rachel Elizabeth Bragg, and M. Seliens, 'Green Exercise in the UK Countryside', *Journal of Environmental Planning and Management* 211 (2007): 50.
5 See inter alia: Jo Barton, R. Hine, and Jules Pretty, 'The Health Benefits of Walking in Greenspaces of High Natural and Heritage Value', *Journal of Integrative Environmental Sciences* 6 (2009): 261–278; E.A. McMahon and D. Estes, 'The Effect of Contact with Natural Environments on Positive and Negative Affect: A Meta-Analysis', *Journal of Positive Psychology* 10 (2015): 507–519.
6 Roger S. Ulrich, 'Visual Landscapes and Physiological Well-Being', *Landscape Research* 4 (2007): 17–23; Jules Pretty et al. 'The Mental and Physical Health Outcomes of Green Exercise', *International Journal of Environmental Health Research* 15 (2005): 319–337.
7 Natural Capital Committee, *The State of Natural Capital: Protecting and Improving Natural Capital for Prosperity and Wellbeing*. Third State of Natural Capital Report (2015) Available at: https://www.gov.uk/government/publications/natural-capital-committees-third-state-of-natural-capital-report-government-response
8 'The Great British Sell Off' (Locality, June 2018) Available at: https://locality.org.uk/wp-content/uploads/2018/06/The-Great-British-Sell-Off-FINAL.pdf
9 *Environmental Improvement Plan 2023: first revision of the 25 Year Environment Plan* (Department of Environment Food and Rural Affairs, January 2023). See Goal 10: "Enhancing Beauty, Heritage and Engagement with the Natural Environment" at pp. 243ff. Available at: https://assets.publishing.service.gov.uk/government/uploads/system/uploads/attachment_data/file/1133967/environmental-improvement-plan-2023.pdf
10 The Town Moor Act 1774 (14 Geo. 3 c cv).
11 See, respectively, the Newcastle Upon Tyne Town Moor Act 1988, the City of Norwich Mousehold Heath Scheme Confirmation Act 1884 and City of Norwich Act 1984 and the Clifton and Durdham Downs Act 1861.
12 See, for example, Antonia Layard, 'Parks in the Time of Covid', Available at: https://legalresearch.blogs.bris.ac.uk/2020/04/parks-in-the-time-of-covid-19/
13 Registered common land is "access land" for the purposes of the right to roam under the Countryside etc. Act 2000, s.1 and Sched. 1.
14 For example, in the case of Town Moor in Newcastle upon Tyne by the Newcastle upon Tyne Town Moor Act 1988, s.8.
15 National Planning Policy Framework (2018) paras 99 and 100 discussed further below.
16 See Chapter 1, p. 9–10.
17 See Michel de Certeau, *The Practice of Everyday Life* (Berkeley: University of California Press, 1984), p. 117.
18 Commons Act 2006, s 15.
19 Prescription Act 1832, s.1 . But note the impact of the Supreme Court decision in R (Littlejohns) v Devon CC [2016] EWCA Civ 446 on the ability to register a common right in these circumstances.
20 See *Newhaven Properties and Port Ltd. v East Sussex County Council* (2013) EWCA Civ 2013 (a tidal portion of a beach, covered by water for part of each day, was held to be in principle registrable as a TVG); *R (Beresford) v Sunderland City Council* [2014]1 AC 889 (sports arena maintained by the local authority, with seating etc. provided); *R (Lewis) v Redcar Borough Council* [2010] UKSC 11 (golf course over which public also exercised recreational use).
21 *Oxfordshire County Council v Oxford City Council and Robinson* (2006) UKHL 25.
22 *R v Oxfordshire County Council ex parte Sunningwell Parish Council* (2000) 1 AC 335

23 Growth and Infrastructure Act 2013, section 16, inserting a new Section 15 C and Schedule 1A to the Commons Act 2006 defining "trigger" events barring applications to register land as a TVG.

24 Commons Act 2006, s.6 (1).

25 That is, under s.1 and Schedule 1, Countryside and Rights of Way Act 2000.

26 *R (Lancashire County Council) v Secretary of State for the Environment Food and Rural Affairs; R (NHS Property Services Ltd.) v Surrey County Council* [2019] UKSC 58 (11 December 2019).

27 *R (Newhaven Port and Properties Ltd v East Sussex County Council* (2015) UKSC 7

28 In the Lancaster case, this was the Education Act 1944, ss.8; Education Act 1996, ss. 13 and 14; Education Act 2002, s. 75 (4); School Premises Regulations 2012, SI 2012/1943, reg. 10.

29 (2019) UKSC 58 para 56 et seq.

30 See excerpts from planning inspectors report at (2019) UKSC 58 para 15 (Lord Carnwath).

31 As noted above, registered common land is "access land" under the Countryside and Rights of Way Act 2000.

32 See further Christopher Rodgers and Duncan Mackay, "Creating 'New' Commons for the Twenty First Century: Innovative Legal Models for 'Green Space'", *Journal of Environmental Planning and Management* 61 (2018): 1051–1069. DOI:10/1080/09640568.2017.1333407. And *New Commons Toolkit*, available at: https://www.ncl.ac.uk/media/wwwnclacuk/newcastleuniversitylawschool/files/Rodgers_Commons%20Toolkit_with%20forms%20(22%20April).pdf

33 This is already happening in some cases. For example, cities like Birmingham that are part of the biophilic cities network have commitments in their plans, as does London's proposed National Park City. See the National Park City Partnership, *Draft Charter for a Greater London National Park City* (2015), www.nationalparkcity.london; and (for Birmingham) *BiophllicCities* (2016) available at: http://biophiliccities.org/?s=birmingham

34 National Planning Policy Framework (Department for Communities and Local Government, 2018) para 100 ("NPPF").

35 Ibid., paras 99 and 100 (c).

Legal Tables

Note: Page numbers followed by "n" denote endnotes

Subject Index

Note: *Italic* page numbers refer to figures and page numbers followed by "n" denote endnotes.

For Product Safety Concerns and Information please contact our EU representative GPSR@taylorandfrancis.com
Taylor & Francis Verlag GmbH, Kaufingerstraße 24, 80331 München, Germany

www.ingramcontent.com/pod-product-compliance
Ingram Content Group UK Ltd.
Pitfield, Milton Keynes, MK11 3LW, UK
UKHW020147200926
475702UK00016B/2600

* 9 7 8 1 0 3 2 0 6 9 2 1 0 *